Tsunami
in a Time of War

Tsunami in a Time of War:
Aid, Activism & Reconstruction in Sri Lanka & Aceh

(A Special Double Issue of *Domains*)

Malathi de Alwis &
Eva-Lotta Hedman
(editors)

ICES, COLOMBO

Tsunami in a Time War: Aid, Activism & Reconstruction in
Sri Lanka & Aceh
Malathi de Alwis & Eva-Lotta Hedman (ed.)

ISBN: 978-0-9748839-5-3

This volume is also a special double issue of *Domains:*
The Journal of the International Centre of Ethnic Studies (no. 4 & 5)
(ISSN: 1391-9768).

Cover photograph by Pradeep Jeganathan of mural detail from Tsunami
Memorial, Peraliya, Sri Lanka.

•

This publication reports on a research project financed by
Canada's International Development Research Centre
(www.idrc.ca).

IDRC ✳ CRDI   Canadä

CONTENTS

Sri Lanka

Palk Strait

Jaffna

Mannar

Vavuniya

Trincomalee

Puttalam

Batticaloa

Amparai

Colombo

Indian Ocean

Galle

Matara

0   50   100 km

Map of Aceh, Indonesia

Introduction

Malathi de Alwis & Eva-Lotta Hedman

On December 26th, 2004, an earthquake off the west coast of Northern Sumatra triggered a tsunami of unprecedented proportions which decimated much of the coastal regions of Aceh Province in Indonesia. It flattened entire villages and left an estimated 164,000 dead and 500,000 displaced, in its wake. The second worst affected country was Sri Lanka with three fourths of its coastline devastated, an estimated 36,000 dead and over a million displaced. In both countries, three times more women died than men and more than a third of the victims were children, thus irrevocably transforming family structures and kin networks; in some instances, entire families were wiped out and it was not uncommon to hear people saying that they had lost 30 sometimes 56 or 84 relatives, within 15 minutes. The acute grief, mental trauma and daily suffering the tsunami engendered defies even these staggering series of enumeration and computation.

In both Aceh and Sri Lanka, the devastation of the tsunami and rehabilitation and reconstruction in its aftermath, also intersected with and shaped the politics of ongoing civil wars leading to militant groups and government forces launching new battles over control of people, land, livelihoods and humanitarian/development aid. In addition, an extremely large influx of new International Non-Governmental Organizations (INGOs) further de-stabilized the precarious balance among central and regional state authorities, militant groups, and in the case of Sri Lanka, already existing United Nations (UN), multi-lateral and bi-lateral donors and other international humanitarian and development organizations providing a variety of services to those displaced/affected by the conflict. Aceh, on the other hand, was re-encountering an INGO/UN presence after over a year's hiatus, since most INGO and UN organizations had been made to leave after the declaration of a state of emergency in Aceh and the re-commencement of hostilities between the Free Aceh Movement or GAM (*Gerakan Aceh Merdeka*) and the Government of Indonesia (GoI), in May 2003.

Indeed, it is this INGO/UN presence and pressure which is primarily credited with pushing the Indonesian government to lift the civil emergency in Aceh, in June 2005, and most importantly, enabling the Helsinki-brokered peace talks which led to the signing of a Memorandum of Understanding to end the conflict between GAM and the GoI , on August 15th, 2005. In fact, politics in Aceh has experienced a sea change with the holding of local elections in December 2006 and GAM joining mainstream politics –GAM candidate Mr. Irwandi Yusuf was elected Governor of Aceh with additional landslide victories for GAM in fifteen out of nineteen districts. In Sri Lanka, in contrast, a tenuous ceasefire, between the LTTE (Liberation Tigers of Tamil Eelam) and the Government of Sri Lanka (GoSL), which had been in place since February 2002, broke down soon after the tsunami around the issue of the disbursement of tsunami aid. Fighting has since intensified between the LTTE and the Government of Sri Lanka with defense forces pushing back the LTTE from the Eastern Province (with the help of the LTTE breakaway faction, the Karuna group or TMVP), in July 2007, and currently also attempting to rout them from their last stronghold in the Northern Province, as this issue goes to press.

This marked divergence in the political trajectories of these two countries in the aftermath of the December 2004 tsunami, provided the overarching framework for this 2-year, multi-sited project which was funded by the International Development Research Centre (IDRC)/Centre de recherches pour le developpement international (CRDI) of Canada and administered by the International Centre for Ethnic Studies, Colombo. Our working hypothesis was that what was at stake in these different political outcomes was the differential articulation of parallel or competing humanitarian aid delivery systems in Sri Lanka and Indonesia, which were also regimes of governmentality deployed through practices of the state, counter-state, and I/NGOs/UN/multi and bi-lateral donors.

In fact, in the post-tsunami period, INGOs became one of the most crucial conduits of aid and administrative functions in these two zones. In the Sri Lankan case, the LTTE counter-state paralleled the Sri Lankan government's infrastructure with its own police force, law courts, schools, banks, tax system and postal service in the areas under its control. It had no real counterpart, in the Indonesian case, as GAM did not exercise territorial control in such ways as to allow it to govern over a population within a demarcated militarised area. Despite this marked difference, the presence and practices of INGOs in both regions demonstrated considerable similarities. Nonetheless, the articulation of INGO programs with state and rebel regimes was distinct in each situation; in Sri Lanka, the LTTE wished to compete with the government of Sri Lanka to partner with INGO-led programs, or better still, lead its own donor-funded humanitarian programs. In contrast, GAM successfully negotiated for some of its members to be appointed to the key Indonesian government agency concerned with the rehabilitation and reconstruction of Aceh, the BRR (*Badan Rehabilitasi dan Rekonstruksi* or Rehabilitation and Reconstruction Agency for Nanggroe Aceh Darusalam and Nias). This difference is crucial to understanding the dynamics of conflict exacerbation in Sri Lanka and its abatement in Aceh in the context of the post-tsunami proliferation of humanitarian interventions.

It was thus considered critical that this research project should examine the articulation of aid programs with state and non-state regimes, from the point of view of the affected communities: How are material humanitarian aid programs, such as infrastructural reconstruction and livelihood rehabilitation, as well as non-material aid programs, such as psychosocial interventions, actualized in the lived world? Do they exacerbate or alleviate existing fissures between and within communities? Are new fissures created? And if so, on what fault lines?

While these were guiding questions which were carried to the field by all the researchers on the project, we also did not wish the researchers to be circumscribed by them as we wanted each field study to benefit from that researcher's knowledge and experience of the affected regions given their scholarly and/or activist involvement in those regions. It was thus very clear from quite early on in the project that the focus would not be on merely documenting and analysing what was currently taking place in the field in conjunction with post-tsunami rehabilitation and reconstruction, but that communities, contexts and practices would also be historicized and attempts made to provide a genealogy of some of the crucial categories being mobilised in the field by INGOs involved in humanitarian rehabilitation, reconstruction and development.

This stress on the individual researcher's knowledge, research/activist experience with particular communities and regions as well as his/her disciplinary training, also enabled us to take a more flexible and innovative approach to exploring the effects of humanitarian interventions in the context of (former) protracted militarised conflict, while also benefiting from the comparative insights and questions raised in the course of our collaborative research project. Rather than attempting a one-on-one comparison between Sri Lanka and Aceh Province, say for example, asking a researcher doing fieldwork in Aceh and a researcher doing fieldwork in Sri Lanka to study a similar topic such as people's responses to new housing settlements, we encouraged researchers to analyse and write about what they felt most passionate about in terms of what they were encountering in the field and also what debates – be it within the specific community or group with whom they were working or the INGOs/state/militant group with whom they were engaging – in which they could intervene in the most productive manner. Even when two researchers, working in two different regions, focused on a similar research topic, i.e., post-tsunami protest movements, the very distinctive and different ways in which they approached as well as analysed their subject matter produced remarkably dissimilar papers which nevertheless resonated with each other in interesting ways. The end results we share with you here thus have gained much in nuance and depth due to this flexible approach while not precluding stimulating conversations, not always explicit, across disciplines, provinces and countries.

Research Methodologies

The fact that we were working with a multi-disciplinary team of researchers –3 socio-cultural anthropologists, 2 political scientists, a human geographer, an urban planner and a demographer –meant that the project, as a whole, has benefited from a richness and variety of research methodologies. While anthropological tools such as 'participant observation,' with researchers and their research assistants living with and participating in the everyday activities of the communities they were studying, predominated, other methodologies such as formal interviews, focus group discussions, surveys and media monitoring were also mobilized. The fact that this project extended over a period of two years and was often conducted in areas where researchers had lived and worked prior to the tsunami, and in countries where some of the researchers were actually domiciled, helped immensely in building relationships of trust and mutual support as well as enabling a much deeper and more contextualized understanding and analysis of the situation on the ground. These more micro-level methodologies were also supplemented by some researchers who were also involved in macro-level analyses of broader global and national trajectories of aid and governance that impinge on localized communities.

Researchers constantly updated each other on their ongoing research findings and were open to re-evaluating their research questions and methodology based on such discussions. One significant outcome of such discussion and reflection was the realization that in order to better understand the extremely diverse aid scenarios in Sri Lanka based on whether the community under study was located in a 'war zone', a 'border' region or neither, whether it was controlled by state troops or LTTE or the rebel faction TMVP or all three in varying degrees, whether it was a predominantly Tamil or Muslim or Sinhala area, some of the researchers needed to focus on more than one field site in order to get a better sense of this variation – sometimes within one province, sometimes across provinces. This also resulted in the unusual phenomenon of two of the researchers deciding to conduct a joint research study in a rather isolated region at the extreme edge of the Southern Province which suddenly came under attack by the LTTE, mid-way through the research project.

This constantly shifting political scenario in Sri Lanka also resulted in the restriction and re-consideration of certain research sites and methodologies. For example, plans to conduct research in LTTE-controlled areas, in the Northern Province, had to be jettisoned, once those regions started becoming more and more volatile as the ceasefire started breaking down, in mid 2005. Researchers working in the Eastern Province had to cut back on travel during the Sri Lankan government's battle to 're-secure' previously LTTE-controlled areas in that province or when certain routes into the province became the target of LTTE claymore mine and suicide bomb attacks. Even the seemingly innocuous act of photographing memorials to those who had died in the tsunami and during the conflict, in the Eastern Province, had to be re-thought in light of increased surveillance by government armed forces regarding certain monuments which had been constructed by militant groups. Similarly in Aceh, public comments made about GAM as well as the Indonesian government, based on research findings, came under undue scrutiny by both parties concerned.

Contours of Conflict
There now exists an extensive literature on the conflicts in both Sri Lanka and Aceh but we shall attempt to briefly lay out some of its more current contours given its intersection with our research and me-

thods. In Sri Lanka, Tamil militants, now primarily the LTTE,[1] have been waging a separatist war against the Sri Lankan state, for the past thirty years. Sri Lanka's present is thus an expression of a long history and geography of struggle well-documented by Sri Lankan and Sri Lankanist scholars.[2] The conflict has spawned large-scale displacement within the country and well beyond its borders, where a significant Sri Lankan Tamil diaspora has emerged.[3] The death toll resulting from this protracted war will soon exceed a million.

The signing of the most recent and to date longest-running Ceasefire Agreement between the GoSL and the LTTE, with the facilitation of the Norwegian government, in February 2002, resulted in enabling the economy to be rebuilt and the LTTE to engage with the GoSL and wider society as a legitimate political entity. However, the escalation of Ceasefire violations, despite the presence of a Scandinavian-led international monitoring body –the Sri Lanka Monitoring Mission (SLMM) — primarily by the LTTE and its breakaway group,

---

[1] The LTTE killed off several leaders and many members of other Tamil militant groups, during the late 1980s. For a detailed account of the rise of the LTTE, see M. R Narayan Swamy, *Tigers of Lanka: From Boys to Guerillas* (Delhi: Konark Publishers, 1994).

[2] See for example, Charles Abeysekera and Newton Gunasinghe (eds.). *Facets of Ethnicity in Sri Lanka* (Colombo: Social Scientists' Association, 1987), Jonathan Spencer (ed.), *Sri Lanka: history and the roots of conflict* (London/New York, Routledge, 1990). Pradeep Jeganathan and Qadri Ismail (eds), *Unmaking the Nation: The Politics of Identity and History in Modern Sri Lanka* (Colombo, Social Scientists' Association, [1995] 2009), Neelan Tiruchelvam, "Sri Lanka's Ethnic Conflict and Preventive Action: The Role of NGOs" in, ed., R. Rotberg, *Vigilance and Vengeance: NGOs Preventing Ethnic Conflict in Divided Societies* (Washington DC, Brookings Institution, 1996), pp. 147-164 and Rajan Hoole, *Sri Lanka: The Arrogance of Power – Myths, Decadence & Murder* (Sri Lanka: University Teachers for Human Rights [Jaffna], 2001).

[3] See for example, Rohini Hensman, *Journey Without a Destination* (Colombo: Centre for Society and Religion, 1993), Valentine Daniel "Suffering Nation and Alienation" in, eds., Arthur Kleinman, Veena Das and Margaret Lock, *Social Suffering* (Berkeley, University of California Press, 1997), pp. 309-58 and Oivind Fuglerud, *Life on the Outside: The Tamil Diaspora and Long Distance Nationalism* (London: Pluto Press, 1999).

the Karuna faction or TMVP (supported by the Sri Lankan government, first clandestinely and then openly), the LTTE's continued conscription of children, assassinations and abductions of rivals and their intransigence over their ISGA (Interim Self-governing Authority) proposals, had greatly undermined the stability of the ceasefire, when the tsunami occurred.

Indeed, it was the opinion of many political analysts that the destruction wrought on LTTE cadres, camps and weapons, by the tsunami, had forestalled a planned, mid-January UDI (Unilateral Declaration of Independence) or military offensive by the LTTE. Though the GoSL and the LTTE did cooperate significantly during emergency and rehabilitative operations in the immediate aftermath of the tsunami, relations began to sour over disagreements surrounding the framing and (non) execution of the Post-Tsunami Operating Mechanism (P-TOM), a humanitarian assistance disbursement agreement which was signed between the GoSL and the LTTE. An undeclared war thus reigned from around August 2006 until the GoSL formally declared the Ceasefire Agreement null and void, on January 2nd, 2008.

During this period, the GoSL, with the support of the Karuna faction, began a systematic operation to rout the LTTE from the Eastern Province, finally declaring victory on July 18th, 2007. Over 150,000 people, primarily Tamils and Muslims, were displaced with many having to flee their newly-rebuilt tsunami houses, scores of which were seriously damaged by shelling and aerial bombardment. Subsequently, the GoSL launched a massive re-building programme entitled the Eastern Re-awakening (*Nagenahira Navodaya*) and strongly urged INGOs, the UN and multi and bi-lateral donors to participate in this programme. In March 2008, local government elections were held in Batticaloa and in May 2008, this was extended to holding Eastern Provincial Council elections, both of which were won decisively by the TMVP (amidst charges of voter intimidation) which jointly contested some of the seats with the GoSL. In August 2008, the GoSL focused its attention on routing the LTTE from the Northern Province. As we go to press, GoSL forces have captured all key towns previously under

LTTE control, confining them to a 45 square kilometre strip of land on the north eastern coast of Sri Lanka. This has also led to a humanitarian crisis of unprecedented proportions as around 200,000 civilians are trapped in this area, along with the LTTE; they are vulnerable to shelling by the GoSL while being used as human shields by the LTTE who are shooting those who attempt to escape areas under their control on the grounds that they will only allow civilians safe passage into GoSL areas once the GoSL signs a ceasefire agreement with them. Such an agreement has been refused by the GoSL which is arguing that this is only a ploy for the LTTE to re-arm themselves and are thus demanding that they surrender.

On the northern tip of the island of Sumatra in Indonesia, Aceh has seen intermittent, militarized conflict between the 1976 inception of a local armed opposition mobilized in the name of 'Aceh Merdeka' (Free Aceh), and the 2005 implementation of the internationally-monitored peace accords. In the aftermath of the demobilization of rebel and government troops, and the wider demilitarization of politics and society in the province, Acehnese voters went to the polls and elected, by an overwhelming majority, candidates running on a ticket associated with the (former) armed opposition in fifteen out of nineteen districts. They also elected the former rebel movement-liaison to the international peace monitoring mission (AMM, [Aceh Monitoring Mission]), Irwandi Yusuf, to the post of provincial governor.

As scholars have shown, Aceh traces a distinctive history of mobilisation and struggle, against the advances of colonialism in the 'Aceh War'(1873-1910), against the old ruling class, the *uleebelang*, in the social revolution (1945-1949) that followed the Japanese occupation during World War II, and, against the nature and direction of the new independent Indonesian Republic in the Darul Islam rebellion (1953-1961).[4] The first rebellion in the name of 'Aceh Merdeka' (1976-

---

[4] See, for example, Anthony Reid, *The Contest for North Sumatra: Atjeh, the Netherlands, and Britain, 1858-1898* (Kuala Lumpur: Oxford University Press, 1969), and *The Blood of the People: Revolution and the End of Traditional Rule in Northern Suma-*

1979) remained very limited in scale and scope, with conflict-related displacement confined to leading figures (including Hasan di Tiro who sought refuge in Sweden in 1979, and supporters who escaped into exile in Malaysia).[5] By contrast, the resurfacing of a comparatively miniscule armed movement in 1989 saw the government deploy some six thousand territorial troops for counter-insurgency operations in Aceh, including two battalions of Kopassus and other elite counter-insurgency units. This substantial deployment of security troops anticipated the introduction of stepped-up surveillance, checkpoints, curfews, house raids, and arrests, as well as the further elaboration of methods employed elsewhere, such as the "fence of legs," "sweepings," and "mysterious killings" as Aceh emerged, alongside Papua and East Timor, as yet another officially designated "trouble spot" (daerah rawan) under the (late) New Order regime in Indonesia.[6] This proscribed "shock therapy" treatment for the second uprising also saw a marked increase in conflict-related displaced, with especially young men from

tra (Kuala Lumpur: Oxford University Press, 1979); James T. Siegel, *The Rope of God* (Ann Arbour, MI: University of Michigan, 2000ed.); and C. Van Dijk, *Rebellion under the Banner of Islam: The Darul Islam Rebellion in Indonesia* (The Hague: Martinus Nijhoff, 1981).

[5] On the emergence of Aceh Merdeka in 1976, see Nazaruddin Sjamsuddin, "Issues and Politics of Regionalism in Indonesia: Evaluating the Acehnese Experience," in *Armed Separatism in Southeast Asia*, ed., Lim Joo-Jock with Vani S. (Singapore: ISEAS, 1984), pp. 111-128; Tim Kell, *The Roots of Acehnese Rebellion, 1989-1992* (Ithaca, NY: Cornell Modern Indonesia Project, 1995); and Geoffrey Robinson, "*Rawan* Is as *Rawan* Does: The Origins of Disorder in New Order Aceh," *Indonesia* 66 (October 1998): pp. 127-156 [reprinted in *Violence and the State in Suharto's Indonesia*, ed., Benedict R.O.G. Anderson (Ithaca, NY: Cornell Southeast Asia Program Publications, 2001), pp. 213-242].

[6] Amnesty International, *Shock Therapy: Restoring Order in Aceh, 1989-1993* (London: Amnesty International, 1993). For a more recent account of the introduction of military operations code-named "*Jaring Merah*," or Red Net, and remembered in Aceh as "DOM" (*"Daerah Operasi Militer"*), see also Rizal Sukma, *Security Operations in Aceh: Goals, Consequences, and Lessons* (Washington, DC: East West Center Policy Studies 3, 2004), pp. 8-11.

affected villages seeking refuge across the Malacca Straits, in Malaysia in the early 1990s.[7]

In the context of wider processes of democratization, demilitarization and decentralization underway in post-Suharto Indonesia, the dynamics of conflict, violence and displacement in Aceh deepened further after 1998.[8] First of all, entire communities moved en masse, "sometimes in public and politicized ways," in the face of (feared or rumoured) violence, thus introducing the notion of *eksodus* (exodus) into local political discourse as government troops sought to regain control of villages in May- June 1999.[9] Moreover, the declaration of martial law in May 2003 introduced some 40,000 Indonesian government troops into Aceh, thus further militarizing the conflict. Under the rubric of 'Operasi Terpadu' (Integrated Operation), forced displacement took on added significance as the Indonesian government for the first time publicly announced plans for the mass evacuation of civilian populations to form part of counter-insurgency operations in Aceh.[10]

---

[7] See, for example, Diana Wong and Teuku Afrisal, "Political Violence and Migration: Recent Acehnese Migration to Malaysia," in *Three Papers on Indonesia and Displacement* (Jakarta: Ford Foundation, June 2002).

[8] The following works offer illuminating, and contrasting perspectives on this more recent period: Matthew N. Davies, *Indonesia's War over Aceh: Last Stand on Mecca's Porch* (London: Routledge, 2006); John Martinkus, *Indonesia's Secret War in Aceh* (Sydney: Random House Australia, 2004); and Anthony Reid (ed.), *Verandah of Violence: The Background to the Aceh Problem* (Singapore: Singapore University Press, 2006).

[9] See, for example, Edward Aspinall, "Place and Displacement in the Aceh Conflict," in , ed., Eva-Lotta E. Hedman *Conflict, Violence and Displacement in Indonesia* (Cornell University, Ithaca, NY: Cornell Southeast Asia Program Publications, 2008), p. 129 - 131.

[10] See, for example, Eva-Lotta E. Hedman, "A State of Emergency, A Strategy of War: Internal Displacement, Forced Relocation, and Involuntary Return in Aceh, in , ed., Hedman , *Aceh under Martial Law: Conflict, Violence and Displacement* , Refugee Studies Centre Working Paper, no. 24 (University of Oxford: Refugee Studies Centre, - 2005), and the other essays in this Working Paper.

As noted above, the post-tsunami, Helsinki-brokered peace talks and subsequent Memorandum of Understanding signed between the GoI and GAM, in August 2005, has augured a marked transformation in Acehnese socio-political life. The de-militarization of Aceh, with the demobilization and disarmament of GAM and the reduction and relocation of the Indonesian National Army (TNI), was overseen by an European Union-led international observer mission – the AMM, from August 2005-December 2006. In this context, the BRA (Aceh Peace-Reintegration Agency), was established in February 2006, with government funding to establish reintegration programs whose beneficiaries would include (former) militia groups, GAM supporters who had surrendered prior to the MoU, and other "conflict-affected persons" throughout all rural communities. BRA continues to be operational today with offices at the district level in several parts of Aceh. However, as it is dependent on funding from the central government, which is often delayed and criticised as being rather meagre, the reintegration work BRA is involved in tends to be rather slow and not very effective.

Despite making such major strides in autonomous governance, after the August 2006 elections, the provincial government continues to have problems wresting guarantees from the GoI that they will honour the recently passed Aceh Administration Law (2006) which finally contains provisions which enables the province to retain a significant percentage of the oil and gas revenues which the GoI earns from the province. The provincial government also continues to be at logger heads with BRR (the Aceh-Nias Rehabilitation and Reconstruction Agency) which is seen as an arm of the GoI though it has many Acehnese and former GAM rebels working in its offices. Charges of corruption and mismanagement also continue to be hurled at BRR which is planning to close its offices in April 2009.

Continued peace in Aceh will very much depend on the processes of democratization which have already been set in place in Aceh, and will hopefully be further concretised through legislative elections which will take place in April 2009. It is clear that economic rein-

tegration must be paralleled with genuine political integration and devolution of power, not just in Aceh but in also the newly-'liberated'
areas of Sri Lanka.

Critical Interventions

One of the significant features of this special issue is that many
of the papers have attempted to offer a critical analysis, sometimes even
a genealogical exploration, of several key categories mobilised in humanitarian and development aid discourses and practices. The value addition, as it were, is that such analyses and explorations are concretised in
and through ethnographic examples and case studies. Our opening essay by Jennifer Hyndman explores the relatively new, neo-liberal concept – aid effectiveness – within 'foreign aid' discourses. Aid
effectiveness, notes Hyndman, "aims to utilise international assistance
most efficiently by eliminating countries with protectionist economic
policies or corrupt, unstable governments from the recipient list".
Though this policy seeks to emphasise government-to-government
partnerships and 'ownership' by host governments, it is nevertheless an
agenda which is driven by donors rather than host governments. However, Hyndman's analysis of CIDA's (Canadian International Development Agency) deployment of this policy reveals how the tsunami was
one instance where a donor government's agenda was forestalled by
other forces. Hyndman also usefully historicizes the unfolding of 'foreign aid' in Sri Lanka, from the 1970s, thus enabling us to better understand its more current incarnation post-tsunami.

Pradeep Jeganathan, on the other hand, offers a fascinating genealogy of a rather hoary old term within aid discourses – 'community.'
His extrapolation here parallels in interesting ways, Malathi de Alwis'
unpacking of another much used and abused term within development
aid discourses and practices – 'participation'. However, while de Alwis'
analysis of 'participation,' and also 'empowerment,' performs more of a
prefacatory role for her interrogation of the 'political,' Jeganathan's
more sustained unraveling of 'community,' over the longue duree, offers us insight into how a term which has its origins in Orientalist, mis-

sionary and colonial knowledge is appropriated and deployed within a post colonial context, prior to the advent of INGOs. This original and thought provoking extrapolation addresses a *lacunae* in the pioneering work of development anthropologists such as Roderick Stirrat who is invoked by both Jeganathan and de Alwis. Jeganathan concludes with some revealing examples of the post tsunami manifestation of 'community' in INGO-suffused, 'rural' Sri Lanka.

Paralleling Jeganathan's concern, Saiful Mahdi also turns his attention on how notions of community operate in Aceh. But, rather than taking a genealogical approach, his analysis seeks to prise apart the indigenous term *gampöng*, a spatial and cultural concept of community that had been sidelined by the Javanese term *kelurahan*, particularly in the urban areas of Aceh, through President Suharto's administrative interventions. While the leader of a *gampöng* is elected by those within that *gampong*, the leader of a *kelurahan* is a government appointee and thus unable to garner the respect and cooperation of the community, unlike the former. Noting that the older concept of *gampöng* seems to be undergoing a revival post-tsunami, Mahdi nevertheless wonders whether the meaning of *gampöng* as a 'place' with 'soul' can be sustained in the new, increasingly isolated and isolating, fragmented and fragmenting settlements that have been built in the tsunami's aftermath.

Mahdi's somewhat apprehensive concluding thoughts regarding the often unforeseen consequences of humanitarian aid is addressed more boldly by de Alwis who begins her paper with Mary Douglas' provocative comment that charity wounds. The inequality which exists between givers and takers is an overarching feature of any aid intervention and is sought to be mitigated through humanitarian and development rhetoric and practices which have embraced concepts such as 'participation,' 'capacity building' and 'empowerment' which de Alwis proceeds to problematize, as noted above. While de Alwis' focus is primarily on how these terms are mobilized by humanitarian aid organizations, Eva-Lotta Hedman's discussion of a key category which appears in the UN Guiding Principles – 'IDP' (Internally Displaced Person) –

illuminates how the circulation of this term in post-tsunami Aceh has anticipated the articulation of contested meanings and positions on the part of the tsunami affected, the war affected and the Indonesian government. The Indonesian government has shied away from using this term to categorise those who had been displaced by the tsunami, while those who had been displaced due to militarised conflict has instead sought to articulate their 'right to return' with reference to the Guiding Principles on Internally Displaced Persons. In a separate campaign, others displaced by the tsunami and housed in so-called temporary relocation 'barracks' also invoked a humanitarian needs discourse in protesting new government distinctions introduced among the displaced on grounds of pre-tsunami property relations (e.g., squatters, renters, or owners of residence). As instances or 'cases' of entangled encounters between, on the one hand, the novel forms of governmentality introduced to regulate and improve the lives of 'IDPs' in post-tsunami Aceh and, on the other hand, their intended 'beneficiaries' or subject populations, these campaigns help illuminate the ways in which something new is created, through political contestation and collective action.

While the central focus of Jacqueline Siapno's paper is the vexed relationships between Acehnese women (especially former GAM combatants and sympathizers) and male GAM members as well as Acehnese women and humanitarian aid workers and 'experts,' she also thoughtfully problematizes the concept of 'reconstruction' dwelling particularly on the reconstruction of religious thought, in this instance, Islam. Siapno points to a troubling conundrum which exists in Acehnese society today but nonetheless remains unaddressed, namely, the new Islamism that is allegedly being pushed by the Indonesian government and which primarily plays out through the moral policing of women. The new GAM-led, secular, provincial government chooses to ignore its increasing sway on Acehnese society, particularly post-tsunami, given the now common perception that it was women's 'loose' behaviour which brought upon the tsunami, along with international humanitarian and development aid organizations (with the exception of Muslim aid organizations) which similarly refuse to engage with Islam while simultaneously calling for women's empowerment and 'gender mainstreaming'.

Vivian Choi breaks new ground through her exploration of how technological innovations such as GIS (Geographic Information Systems) have transformed the production and dissemination of knowledge and are now part and parcel of humanitarian interventions. Such technologies have also enabled a shift from 'disaster management' to 'disaster risk management' thus transferring the focus of nation-states to preparedness, reduction of risk and an increased concern with security. By the inclusion of 'terrorist attacks' in the inventory system set up by the new National Disaster Management Centre in Sri Lanka,[11] observes Choi, the Sri Lanka state has integrated natural and man-made disasters as an inevitable part of Sri Lankan life. Does such a formulation lead to a safer Sri Lanka or merely produce a false sense of security?

Another interesting feature of this volume is the complicated and varied array of agents and institutions that are analysed and explored. Hyndman's primarily macro analysis of 'foreign aid' also provides a nuanced reading of a specific bi-lateral donor, CIDA, which offers an important counterpoint to some of the other papers which have, in most instances, homogenized humanitarian aid organizations under the general catch-all of INGOs. Jeganathan's careful highlighting of specific INGO rhetoric and practices however, is an exception here. Choi, performs a similarly important task in de-homogenizing the state by focusing our attention on the workings of a specific state institution, the newly-formed National Disaster Management Centre. She also seeks to bridge the gap between the work of the state and those on whom it impinges by offering us some 'grounded,' localized perceptions of the tsunami warning tower in Kalmunai recently constructed by the Centre.

Mahdi, de Alwis, Hedman and Siapno, on the other hand, are primarily concerned with specific individuals, communities and

---

[11] This has now been dropped.

movements 'on the ground.' Mahdi's focus on new settlements and Siapno's on gendered interactions and religious 'reconstruction' make important contributions to the growing literature on these topics post-tsunami. In contrast, de Alwis and Hedman focus on a much less studied and analysed phenomenon within the broader field of disaster studies – political protests. De Alwis engages with a post-structuralist reading of the 'political' which foregrounds mutual antagonisms and constantly puts the parameters of the political itself into question. Is the mobilization of such a 'political' no longer possible, she wonders, in a war-weary, post-tsunami Sri Lanka where most public action is now either usurped by opportunistic political parties or NGOs and INGOs which "mistake 'governance' for 'politics'". Hedman, on the other hand, explores the governmentalities of displacement and their effects to show how, in two entangled encounters with their respective 'subject populations', or 'beneficiaries,' such schemes for improvement have served to produce particular nodes of 'friction'. As Hedman's focus on the protest campaigns that emerge out of this friction suggests, they point to the possibility of the formation of new subjectivities through contestation. In the words of Nicholas Rose, "a difference is introduced into history in the form of politics" (see p. 141).

Policy Interventions

A crucial aspect of this project has also been concerned with thinking about productive ways in which we can engage with policy makers to share many of the insights we have gleaned from our two years of research and analysis. Most importantly, we wanted to use this opportunity to think more deeply and critically about the frequent disconnect which exists between research and policy, to interrogate why this occurs and to explore more innovative and alternative ways through which we could work with policy makers to re-think and reflect on some crucial conceptual foundations that we have come to find problematic.

In this regard, we found a very thoughtful and enthusiastic ally in development consultant and economist Sunil Bastian who has many years of experience working both as a researcher as well as a consultant to donors and policy makers and is thus very sensitive to the problems faced by those on both sides of this divide. Bastian has been involved in all our methodological and research discussions, these past two years, and has contributed in important ways to raise many issues concerning policy, at these meetings as well as to enable roundtable discussions among researchers and policy makers/donors/ humanitarian and development workers in both Aceh and Sri Lanka. It is thus fitting that his paper concludes this special issue by drawing on all the research papers as well as bringing many of his own insights to bear on this troubled issue of post-tsunami reconstruction and rehabilitation.

Bastian problematizes the notion of 'emergency' that energizes most humanitarian organisations and demonstrates that managerial tools such as Guiding Principles are no answer to the complex issues they face at the point of intervention. He also engages a number of issues raised by the research studies, such as the importance of constantly questioning the theoretical and methodological models used by humanitarian aid organizations, the need for a better understanding of the nature of the state and its possible role in rehabilitation, the tragic consequences which result when social organization and land tenure patterns are ignored by aid organizations, and that the repetition of the 'community participation' mantra does not ensure equity. Indeed, for policies and projects to be successful, concludes Bastian, it is crucial that implementors should make use of existing knowledge bases and employ people who have the competence and experience to work in these societies; the institutional structures of the agencies must also be flexible to deal with all kinds of social and political complexities which are unique to each country where they work.

•   •

A project of this scope and complexity would not have been possible without the support and enthusiasm of Dr. Navsharan Singh, Senior Program Officer at IDRC/CRDI, Delhi, who oversaw this project. The various forms of institutional support we received was also crucial: At the International Centre for Ethnic Studies, Colombo, Mr. S. Varatharajan and his two accounts assistants, Ms. Shyamalee and Ms. Preethika, shouldered the burden of balancing the budget and juggling reimbursements in a variety of currencies while Ms. Tharanga de Silva took care of the onerous logistics of visas, travel itineraries and hotel bookings for the several conferences we had during the course of this project. The Aceh Institute in Banda Aceh, Indonesia, and Mr. Saiful Mahdi and Ms. Cut Famelia in particular, were invaluable in helping to coordinate our first research conference which was held in Banda Aceh. In Penang, Malaysia, Professor Francis Loh Kok Wah and Ms. Gaik Lan at the Centre for International Studies, Universiti Sains Malaysia (USM), graciously hosted our workshop on research methodologies.

We are indebted to our researchers for the dedication and seriousness with which they worked on this project and did their utmost to stay in touch via our listhost, to attend conferences and to meet deadlines despite demanding workloads, illnesses and personal tragedies. The conversations in which we were all involved, these past two years, have been both stimulating and productive and are sure to continue long past this project coming to a close. We regret that one of our researchers, Dr. Riwanto Tirtosudarmo (Senior Researcher, Research Centre for Society and Culture, Indonesian Institute of Sciences, Jakarta), is not represented in this issue due to difficulties in scheduling his revisions. Ms. Amila Jayamaha sacrificed her Christmas vacation to copy edit the papers. We are very grateful to her.

Our greatest debt however, is to the communities in which we did our fieldwork – those who welcomed us into their small barrack rooms, crowded tents and later, brand new houses, often with already cracking walls and leaking roofs, to share with us their grief, anger, frustration and hope in spite of their suspicion of and exhaustion with journalists, census takers and researchers who have endlessly questioned, enumerated, tabulated, photographed, filmed and otherwise 'documented' them these past four years. This issue is dedicated to all those affected by conflict and the tsunami, in Aceh and Sri Lanka, who struggle – in desperation and in hope – to live with dignity against insurmountable odds.

The Geopolitics of Pre-Tsunami and Post Tsunami Aid to Sri Lanka[1]

Jennifer Hyndman

Abstract

The 2004 tsunami produced enormous loss and destruction, remarkable media attention, and an extraordinary outpouring of international aid. This paper seeks to argue that tsunami aid cannot be adequately understood if we do not contextualise it historically. It is at once distinct from long-term development aid and part and parcel of humanitarian assistance. Nonetheless, the conditions of its giving and the context in which it has been provided are constitutive of its meaning and impact in the case of Sri Lanka. In this regard, the relationship of the 2004 tsunami to aid disbursements and politics in Sri Lanka is of particular interest. Utilising data from interviews with international non-governmental organisations and aid agencies in Colombo, the salient responses of bilateral and international non-governmental agencies are analysed. A case study of one of these donors – the Canadian International Development

[1] Thanks to the International Development and Research Centre (IDRC) and to the Social Sciences and Humanities Research Council of Canada (SSHRC) for their generous funding of this research. Soundarie David provided critical research assistance in Colombo. Thanks also to Sarah Paynter, Alison Mountz, Robert Lidstone, and Sunil Bastian for their helpful comments on earlier drafts. Dr. Mala de Alwis has been a fearless leader and coordinator for the Sri Lankan part of the IDRC-funded project on post-tsunami responses, along with her effervescent collaborator on the Indonesian side, Dr. Lotta Hedman. I thank them both for their editorial insights on this manuscript, and Mala for sharing her news clippings collection with me once again.

Agency (CIDA) – is employed to provide a historicised and detailed analysis of its relationship with Sri Lanka over time. The various impacts of neoliberal aid policies within CIDA, active conflict in Sri Lanka, and the devastation of the tsunami witnessed worldwide are all probed in relation to CIDA's assistance on the ground. The paper presents evidence from research conducted before and after the tsunami to argue that crisis creates exceptionalism. CIDA dramatically changed its neoliberal application of 'aid effectiveness' policy in Sri Lanka in the wake of the tsunami. While CIDA's response was not exemplary of all bilateral foreign aid agencies to Sri Lanka, it illustrates the historicised and geopolitical antecedents that shape aid practices on the ground.

Foreign aid now seems to be concerned with the total transformation of Sri Lankan society.
– Sunil Bastian[2]

When asked why the 2004 tsunami in the Indian Ocean Basin attracted so much more attention than the devastating Pakistan earthquake that occurred eight months later in 2005, one international aid agency official in Colombo remarked that the magnitude of the tsunami was unprecedented.[3] "This disaster was also very much a media opportunity where visuals of women caught up on trees, the plight of children [and such] were shown so freely. It was almost as if *destruction came to the sitting room.*"[4]

International aid agencies found it easier to provide a multitude of boats to replace those lost by fishers in the wake of the tsunami (see figure 1) than to assess changes in social relations generated by the dis-

---

[2] Sunil Bastian, *The Politics of Foreign Aid in Sri Lanka* (Colombo: International Centre for Ethnic Studies, 2007).

[3] Carl Grundy-Warr and James D. Sidaway, "Political Geographies of Silence and Erasure," *Political Geography*, 25 (2005): 479-481.

[4] Interview #105 with aid agency official, Colombo, Sri Lanka, 2007; emphasis added.

aster. For example, the meaning of widowhood and access to remarriage for those who lost spouses in the tsunami emerged as very different from those who lost spouses to the conflict.[5] Likewise, tallying the amount of foreign aid from various sources allocated to different sectors is far easier than decipherin the dynamics of aid in the contexts of conflict and disaster. This paper aims to fill this gap by tracing the salient shifts in international aid since Sri Lankan independence in 1948, focusing on the dynamics of aid inside one bilateral aid organisation, the Canadian International Development Agency (CIDA), and then teasing out the main issues facing donors in the context of ongoing conflict and the wave of post-tsunami funds. Tacking back and forth between the specific aid policies

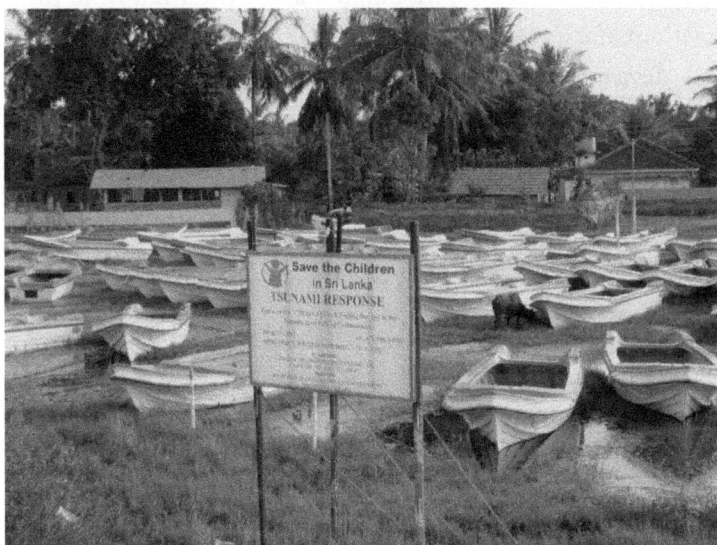

Figure 1: Too many boats, not enough livelihoods,
Batticaloa, Eastern Sri Lanka

---

[5] Jennifer Hyndman, "Feminism, Conflict and Disasters in Post-tsunami Sri Lanka," *Gender, Technology, and Development*, 12, no. 1 (2008): 101-121.

and practices of CIDA and the more general landscapes of aid over time in Sri Lanka allows me to make the argument that aid is a highly geopolitical project that varies among bilateral donors and among international non-governmental organisations (INGOs), UN agencies, and bilateral funders. The agendas and constraints of each shape actions, but policy does not always prevail. Geopolitical considerations are always in play.

A Note on Method

Research for the CIDA case study took place over a period of five years, funded by Canada's Social Sciences and Humanities Research Council (SSHRC). In 2002 and 2005, both pre- and post- tsunami, aid policies and practices in Canada and Sri Lanka were probed.[6] Of course, when the project began in 2002, no one knew that a tsunami would devastate Sri Lanka and rearrange aid relations, but the opportunity to interview aid officials at CIDA both before and after the tsunami in both Ottawa and Colombo led to some unexpected findings presented here.[7] The post-tsunami work of many others has been invaluable to this analysis.[8]

---

[6] Other analyses of these data have been written up elsewhere; see Jennifer Hyndman, "The Securitization of Fear in Post-Tsunami Sri Lanka," *Annals of the Association of American Geographers*, 97, no. 2 (2007): 361-372; elements of the argument presented in part two of this paper are published in Jennifer Hyndman, "Acts of Aid: Neoliberalism in a war zone," *Antipode* (forthcoming).

[7] The shortcomings of such methods are obvious; they are partial snapshots of post-tsunami aid relations. While one can use newspaper articles and other news sources to track political developments, as is done here, temporal and geographical gaps in the research findings remain. Unlike most research conducted in Sri Lanka by this author, the 'fieldwork' for this paper has been based mostly in 'capital cities', speaking with policymakers and aid officials rather than with those for whom the aid may be intended, namely those affected by the twin humanitarian disasters of war and tsunami.

[8] For an analysis of impact of the tsunami on post-tsunami gender relations, for example, see Neloufer de Mel and Kanchana Ruwanpura, *Gendering the Tsunami: Women's Experiences from Sri Lanka* Report Series (Colombo: International Centre

In 2007, the International Development Research Centre (IDRC) sponsored a comparative research project focused specifically on the intersection of war and the 2004 tsunami in Sri Lanka and Aceh, Indonesia. Our part of the research team canvassed all INGOS and bilateral aid agencies operating in Sri Lanka for their analyses of tsunami aid, its relation to conflict, its effectiveness and conflict sensitivity. Interviews were held with willing senior managers at these agencies and lasted from 45 to 80 minutes. Among other responses, tensions, rivalries, and inequities created after the tsunami were recorded.

Outline of the Paper

Tsunami aid cannot be adequately understood if we do not contextualise it historically. It is at once distinct from long-term development aid and part and parcel of humanitarian assistance. Nonetheless, the conditions of its giving and the context in which it has been provided are constitutive of its meaning and impact in the case of Sri Lanka. This paper builds upon a rather brief summary of foreign aid to Sri Lanka since the introduction of neoliberal policies in 1977. The existing literature analysing foreign aid is sizeable, with Sunil Bastian's (2007) book, *The Politics of Foreign Aid in Sri Lanka*, a welcome and comprehensive new tome that elaborates aid politics during this period. After providing a context for foreign aid, the main body of the paper is focused on a sustained analysis of aid from one bilateral donor to Sri Lanka – CIDA. This case study traces Canada's aid to Sri Lanka during the decades of conflict during which many Sri Lankans have sought asylum in Canada. It outlines CIDA's more recent commitment to a one kind of neoliberal development aid strategy, that of 'aid effectiveness', in the context of the war and then of the tsunami. Since acute

for Ethnic Studies, 2006) and Sarala Emmanuel, "Contextualizing Post-Tsunami Challengers: Research Findings of Household Survey in Tsunami Affected Areas," presented in Proceedings of *South Asian Conference on Gender Concerns in Post-Tsunami Reconstruction*, 15-16 July 2005 (Batticaloa: Suriya Women's Development Centre, 2005). On marriage and widowhood, see Jennifer Hyndman, 2008, *op. cit.*

human rights violations have characterised the Sri Lankan landscape, CIDA has delivered its bilateral aid to Sri Lanka in unusual ways, through non-governmental organisations, not government. Since the adoption of aid effectiveness principles in the early 2000s, CIDA developed a plan to wrap up its bilateral aid programming and exit Sri Lanka. The tsunami changed that thinking, and this case study traces why and how.

The last part of the paper probes the relationship of tsunami aid to the conflict, more generally, within the context of Sri Lanka. Using data from the 2007 interviews with INGOs and aid agencies in Colombo, the most salient responses from informants are analysed. While an acute increase in aid after the tsunami is no surprise, evidence suggests that policies and mechanisms related to tsunami aid have exacerbated tensions in three main ways. Inequalities were created or exacerbated in the wake of tsunami aid; the buffer zone policy became a source of tension rather than a public safety provision; and the failure of P-TOMS (the Post-Tsunami Operational Management Structure) was indicative of renewed conflict. Each of these is elaborated.

The Politics of Foreign Aid to Sri Lanka

After its independence in 1948 and during the Cold War, Sri Lanka maintained a strong position as a non-aligned nation in the Third World, following import-substitution policies and eschewing many offers of development aid from the capitalist 'First' World states. In the mid-1970s, the Sri Lankan government unsuccessfully requested loans from the International Monetary Fund (IMF). This changed with the election of the UNP government in 1977 at which time Sri Lanka opened its markets to unbridled capitalist development and export-oriented industrialisation. At the same time, it promised to create a just and free society that included Tamil-speaking people.[9] Within two weeks of taking office, however, police burnt down the public library in Jaffna, sparking a reign of terror and violence against Tamils through-

---

[9] Sri Lanka was the first in South Asia to liberalize its economy in 1977.

out the country.[10] When bankruptcy threatened the newly elected government of J. R. Jayewardene, it opened Sri Lanka's markets to export-oriented industrialisation and thus met the loan conditions of the IMF. Economic liberalisation meant an end to many of the concessions and patronage relations that had kept the peace between otherwise disparate class and ethnic factions, eroding the glue holding ethnic and class alliances together.[11] While foreign dollars began to pour into Sri Lanka, specific segments of the Tamil and Sinhala population remained largely excluded from investment.[12] This cultivated intense resentment among members of the Tamil minority in Sri Lanka's North and Northeast and rural working class Sinhala youth in the South. This new and disparate economic geography was, in part, an outcome of international aid and loan inputs.

According to sociologist Newton Gunasinghe, neoliberal economic policies imposed by the IMF contributed to the rise of militant nationalism and insurgency by the Liberation Tigers of Tamil Eelam (LTTE) and the *Janatha Vimukthi Peramuna* (JVP or People's Liberation Front).[13] Well before the anti-Tamil pogroms of 1977 and 1983, notes Gunasinghe, the introduction of the 1956 *Sinhala Only Act* stirred up communal antagonism and generated considerable resentment among Sri Lankan Tamils.[14] This act declared Sinhala to be the only official language, removing Tamil as one of the state-sanctioned languages. The political controversy generated by the Act was exacerbated by a scheme in the 1970s to standardize examination marks for

---

[10] A. Sivanandan, "Sri Lanka: A Case Study," in *Communities of Resistance: Writing on Black Struggles for Socialism* (London/New York: Verso, 1990), pp. 199-248.

[11] Kristian Stokke, "Sinhalese and Tamil nationalism as post-colonial projects from 'above,' 1948-1983," *Political Geography* 17, no. 1 (1998): 83-113.

[12] Sivanandan, *op. cit.*

[13] Newton Gunasinghe, "Ethnic Conflict in Sri Lanka: Perceptions and Solutions," *Facets of Ethnicity in Sri Lanka* (Colombo: Social Scientists Association, 1987), pp. 61-71.

[14] *Ibid.*

entrance to university, so that those writing exams in the Sinhala medium required fewer marks than those writing in Tamil.[15]

Members of the Sinhalese political elite, through a series of patron-client relations, made general concessions to the Sinhalese rural middle and lower classes in order to be returned to power.[16] From the 1950s to the electoral victory of the United National Party [UNP] in 1977, the government appropriated significant sums from the economy and redistributed some of the surplus through various economic and social development programs.[17] "[T]he system of quotas, permits and licenses, a product of the policy of import substitution made state patronage essential for any individual or commercial venture in the private sector."[18] Economic liberalization meant an end to many of the concessions that held ethnic and class alliances together.[19] Gunasinghe takes this argument a step further and argues that there is a direct link between economic liberalization and the anti-Tamil violence between 1977-1983, and that this violence triggered the civil war between the Sinhalese-dominated state and militant Tamil separatists.[20]

Sri Lanka had become a favourable partner in the eyes of bilateral and multilateral aid and finance institutions due to its commitment to economic deregulation and free market policies and its relatively democratic status in terms of human rights and welfare:

Here indeed was a deal waiting to be made. But it was implicitly a package deal, and one which had visibly to be in place rather quickly. Continued support from the World Bank and IMF would implicitly be contingent on enthusiasm for Sri Lanka in international business circles and in the foreign and trade ministries of the industrialised countries. But foreign capital would look at Sri Lanka again only

---

[15] Sivanandan, *op. cit.*

[16] Stokke (1998), *op. cit.*

[17] Gunasinghe, *op. cit.*

[18] Gunasinghe, *op. cit.*, 199.

[19] Stokke (1998), *op. cit.*

[20] Gunasinghe, *op. cit.*

if the new government could reverse the country's reputation for political radicalism and electoral instability. Hence, a whole series of urgent policy measures to create a new business image....[21]

Distinct from Gunasinghe's argument, Mick Moore refutes any causal link between economic liberalisation and violence, but contends that the large net increase of foreign aid that accompanied open markets, along with a larger and more salient public sector in Sri Lanka, may well have encouraged and sustained authoritarian practices.

This history of foreign aid is also well-documented by Bastian who examines Mark Duffield's thesis in the context of Sri Lanka that aid is a tool in the 'news wars', one for stabilising poorer countries affected by conflict so as to prevent unwanted out-migration from them.[22] Hence, donor countries may use foreign aid as 'soft power' to manage migration flows, not only from poor countries but also from those affected by severe conflict at risk of displacing certain groups beyond their borders. Sri Lanka, I contend, is one such country.

After the 1983 pogroms against Colombo-based Tamils that commonly mark the beginning of armed conflict in Sri Lanka, intense fighting on the battlefield between the LTTE and the Sri Lankan armed forces was exacerbated by acute human rights violations against civilians on the part of the Sri Lankan government as well as the LTTE.[23] Canadian Government officials, among others, visited Sri Lanka and met with the foreign minister of the day to lobby for the cessation of widespread and well-documented abuses against mostly Tamil civilians.[24] The Sri Lankan government did not respond to Ca-

---

[21] Mick Moore, "Economic liberalization versus Political Pluralism in Sri Lanka?" in *Modern Asian Studies* 24, no. 2 (1990): 354-55.

[22] Bastian, 2007, *op. cit.*; Mark Duffield, *Global Governance and the New Wars: The Merging of Development and Security* (London: Zed Books, 2001).

[23] Rajan Hoole, Daya Somasundaram, K. Sritharan and R. Thiranagama, *The Broken Palmyra - The Tamil Crisis in Sri Lanka* (Claremont, CA,: The Sri Lanka Studies Institute, 1990).

[24] Personal communication, Raphael Girard, former Deputy Minister of Immigration, 2004.

nadian, or any other, pleas and continued its unchecked violence against civilians, actions that conditioned Canada's decision to channel all its bilateral aid through non-governmental organisations (NGOs), and not to the government itself. This arrangement was based on the perceived political neutrality of 'civil society' and an assumption that development funds were more secure in the hands of NGOs.

I turn now to an analysis of aid effectiveness principles circulating among a plethora of donors in the global North, then move on to deepen the specific case study of CIDA and illustrate how one bilateral aid agency navigated the conflict in Sri Lanka before the tsunami and how it responded thereafter.

Aid Effectiveness:
From Global Discourse to National Policy

Calls for trade, not aid, have become the watchwords of many developing countries (notably Uganda and Eritrea) as well of INGOs (especially Oxfam). Rich countries spent $280 billion in 2004 subsidising farmers and agribusiness, more than triple the $79 billion they spent on aid. The great irony in the debate between aid and trade is that if all trade barriers were removed and agricultural subsidies eliminated, developing countries would gain $100 billion in annual income according to World Bank estimates, far more than they get in aggregated development aid each year.[25]

During the 1990s, the amount of foreign aid extended to poor countries fell by one third in constant dollars. "It is ironic – and tragic – that just as economic reform has created the best environment in decades for effective assistance, donors have cut back aid sharply."[26] Relative to the economies of donor governments, the proportion of aid fell even further. In the US, overseas development assistance in absolute

---

[25] Celia Dugger, "Trade and Aid to Poorest Seen as Crucial on Agenda for Richest Nations," *The New York Times*, 18 June 2005, p. 8.

[26] Quotation from World Bank cited in J. Cassidy, "Helping Hands: How foreign aid could benefit everybody," *The New Yorker*, 18 March 2005, p. 62.

dollars was halved in a ten year period.[27] This was also true at CIDA where the total budget fell by one third during the 1990s.

CIDA was established in 1968 with a mandate to "reduce poverty and to contribute to a more secure, equitable, and prosperous world."[28] Its budget in 2005-06 was $2.78 billion with another $1 billion in Canadian dollars (CAD) distributed by other government departments and organisations (such as IDRC). For the past eight years, CIDA has adopted a new aid policy shared by a wide range of its peers: aid effectiveness.[29]

Aid effectiveness refers to a popular neoliberal strategy on the part of donors to reward recipient countries with 'sound economic policies' and 'good governance' with international assistance. Identified first by the Organization for the Economic Cooperation and Development (OECD)[30] and promoted by the World Bank,[31] 'strengthening aid effectiveness' is a salient neoliberal policy of development that aims to utilize international assistance most efficiently by eliminating coun-

---

[27] Dramatic budget cuts between 1990 and 1995 saw a decline in CIDA's allocation from cabinet from CAD$3 billion to $1.9 billion. Since the adoption of aid effectiveness policies in the early 2000s, however, CIDA has seen successive increases in its budget of 8% per year, up to a total of more than $3 billion in 2004-05. Not all programs are benefiting equally from these increases: partnership programs [projects coordinated between Canadian NGOs and developing country NGOs] have seen no new money for five years, effectively a reduction, while other areas such as aid effectiveness initiatives have seen more funding; Rick Cameron, Senior Vice-President, CIDA, Address to stakeholders at the Wosk Centre, Simon Fraser University, Vancouver, Canada, 25 November 2004.

[28] CIDA, "About CIDA," CIDA homepage, accessed 18 July 2008 at http://www.acdi-cida.gc.ca/ CIDAWEB/acdicida.nsf/En/NIC-5410529-KFT.

[29] CIDA, "Where is CIDA?," CIDA homepage, accessed 18 July 2008 at http://www.acdi-cida.gc.ca/CIDAWEB/acdicida.nsf/En/JUD-829101441-JQC#2 .

[30] OECD-DAC, "The Paris Declaration," (Development Co-operation Directorate (DCD-DAC), 2008), accessed 8 July 2008 at www.oecd.org/document/18/0,2340,en_2649_3236398_35401554_1_1_1_1,00.html.

[31] World Bank, *Assessing Aid: What Works, What Doesn't and Why* (New York: Oxford University Press, 1998).

tries with protectionist economic policies or corrupt, unstable governments from the recipient list. 'Good governance' and 'sound economic policy' are prerequisites for receiving international aid under this policy rubric. Donors aim to consolidate their aid 'investments' in countries where their 'return' will be greatest in terms of 'pro-poor growth.'[32] The neoclassical economic logic may be simplistic, but has proven politically persuasive to the governments that allocated funds to their aid agencies.[33] International aid budgets have increased where countries have adopted such policies.[34] In short, the allocation of much development aid is being reconfigured both geographically and programmatically.

The neoliberal policy of aid effectiveness overlaps with longstanding geopolitical priorities related to promoting security, democracy, and liberal peace.[35] Threats to national security, in particular, have stoked a geopolitics of fear that less developed countries will 'invade' or infect donor countries if adequate aid transfers are not provided. During the Cold War, foreign aid to developing countries was as much about forging geopolitical allies and proving the superiority of capitalist economies as it was remedying the 'underdevelopment' of non-aligned Third World nations.

The United States, Britain, Canada, the Netherlands, and others have integrated development aid with foreign policy through both security and aid effectiveness lenses. Ultimately, governments in these donor countries argue international cooperation combined with aid effectiveness can make their states more secure.[36] In 2005, the Paris Dec-

---

[32] C. Burnside and D. Dollar, "Aid Policies and Growth. Policy Research Working Paper 1777," (Washington DC: The World Bank, 1997).

[33] J. Knockaert, "Canada making a difference in the world," CIDA presentation and overview, Victoria, Canada, 8 June 2004.

[34] J. Sallot, "Canada discovering foreign aid easier said than done," *The Globe and Mail*, 23 June 2005.

[35] Personal communication, Raphael Girard, former Deputy Minister of Immigration, 2004.

[36] Hyndman, "Acts of Aid: Neoliberalism in a war zone," *op. cit.*

laration – adopted by more than one hundred ministers, heads of agencies and other senior officials of countries and organisations – became evidence of a global commitment to aid effectiveness. The Declaration pledged "to continue to increase efforts in harmonisation, alignment and managing aid for results with a set of monitorable actions and indicators."[37] Aid effectiveness aims to utilise international assistance most efficiently by eliminating countries with protectionist economic policies or corrupt, unstable governments from the recipient list.[38] Donors look for recipient countries with 'good governance' and 'sound economic policy' under this policy rubric.

Aid effectiveness policy emphasises government-to-government partnerships and tries to avoid the hierarchal language of donor-recipient. The language of 'recipient' has been changed to 'host' or 'partner government,' with ownership by the host country as an apparently new feature of this particular aid approach.[39] But Rehman Sobhan argues that the aid effectiveness agenda is actually driven by donors, not host/recipient governments:

> ...donors remained preoccupied with the effectiveness of development assistance in reducing global poverty. The use of the metric of poverty was inspired by the nature of the appeal to those who finance aid – the citizens, largely in the role of taxpayers, in the advanced industrial countries (AIC).... DC

---

[37] The Paris Declaration has been endorsed by more than 100 countries, 26 international organizations, and 14 civil society organizations (S Graves and V. Wheeler, "Good Humanitarian Donorship: Overcoming Obstacles to Improved Collective Donor Performance," Humanitarian Policy Group, discussion paper, December 2006); to read about the indicators and targets for 2010 set out by the Declaration, see, "The Paris Declaration," accessed 18 July 2008 at http://www.oecd.org/document/ 18/ 0,2340,en_2649_ 3236398_ 35401554_1_1_1_1,00.html.

[38] World Bank, *op. cit.*

[39] CIDA, "Strengthening Aid Effectiveness: new approaches to Canada's international assistance," consultation document, February 16 year?; shorter version published June 2001.

[developing country] aid recipients do not vote in AIC elections.[40]

Host governments obviously do not get to select whether they meet the aid effectiveness criteria. The geographical slippage in accountability and semantics is important.

Despite all the talk of 'empowerment', 'partnership' and 'participation', development is still something that is defined and enunciated by the 'first world'. Just as in colonial times, the frameworks and strategies of development are authored outside of the country concerned, grounded in foreign (especially neo-liberal) ideologies and backed up by the long-arm of debt conditionality.[41]

Aid effectiveness rationalises selective aid expenditures by donors

Another common feature of aid effectiveness policy is geographical concentration, or the focus of aid monies to a few select locations for maximum impact. Ian Smillie has raised concerns about the geographical concentration of aid adopted by CIDA and other OECD donors.[42] He cautions against the danger of crowding the 'good performers', by which he means if all donors pursue similar strategies of selectivity in their aid programs, a polarisation of aid funding will occur creating a disparity between 'good' and 'bad' 'partners'. While multilateral coordination of donors might combat this trend to some degree, the principle of selectivity nonetheless threatens to focus aid on 'well-governed' countries, abandoning others to the more meagre assistance provided through the spare change of ad hoc humanitarian aid

---

[40] R. Sobhan, "Aid Effectiveness and Policy Ownership," *Development and Change* 33, no. 3 (2002): 541-542.

[41] C. Mercer, G. Mohan and M. Power, "Towards a critical political geography of African development," *Geoforum* 34, no. 4 (2004): 423.

[42] Ian Smillie, "ODA: Options and Challenges for Canada," commissioned by Canadian Council for International Cooperation, Feb/March 2004; accessed 22 July 2006 at http://www.ccic.ca/e/docs/002_policy_2004-03_oda_options_smillie_report.pdf.

channels. Smillie also argues that geographical concentration cannot be justified in CIDA's case as a way of securing greater leverage in a particular country context. Canada's contributions are simply too small relative to other donors to gain the kind of 'critical mass' and impact alluded to in CIDA's 2001 policy statement endorsing aid effectiveness. At CIDA, however, the policy of strengthening aid effectiveness has been credited with restoring aid allocations which continue to grow at 8% per year.[43]

Does less geographical focus promote more effective aid? "There is no clear, simple link between focus and aid effectiveness …. Dispersion, like focus, needs careful thought and justification,"[44] despite critics who have found that Canadian aid is "scattered…all over the place." In another study, OECD's Development Assistance Committee (OECD-DAC) stated that Canada's aid dispersion was "increasingly problematic", but did not state why beyond the idea that aid impact is diluted.[45]

Aid effectiveness discourse among donors like Canada, Britain, and the Netherlands aims to address poverty to curb the security risks and costly migration effects of underdevelopment.[46] Development aid has become strategic in new ways. "Canadians will benefit from a more prosperous, secure and equitable world. Conversely, the argument goes, Canadians will be less and less secure in a world characterised by pover-

---

[43] Jeff Sallot, "Canada discovering foreign aid easier said than done," *The Globe and Mail*, 23 June 2005.

[44] L. T. Munro, "Focus-Pocus? Thinking Critically about Whether Aid Organizations Should Do Fewer Things in Fewer Countries," *Development and Change*, 36 no. 3 (2005): 425, 427.

[45] OECD-DAC, "Development Co-operation Review – Canada. Organization for Economic Cooperation and Development," Paris 2002; accessed on 26 July 2006 at http://www.oecd.org/document/61/0,2340,en_2649_33721_2409533_1_1_1_1,00.html.

[46] Mark Duffield, "Global Civil War: The Non-Insured, International Containment and Post-Interventionary Society," *Journal of Refugee Studies*, 21 no. 2 (2008):145-165.

ty and unsustainable development." Since "disease does not need a visa," development aid is a tool for enhancing our health and security.[47]

Development assistance is also a political tool for engaging countries that are not 'strategic' to donors in terms of foreign policy or trade.[48] By way of example, "[t]he huge diversion of aid monies into Afghanistan and Iraq is clearly about "stabilizing" areas that are seen to pose a threat to the North."[49] Canada sent more development aid to Afghanistan and Iraq than to all of its other aid recipients put together, pledging $600 million between 2001 and 2007 for Afghanistan alone, a significant sum for a donor with an annual budget of about $3 billion.[50] In contrast, Canada's total commitment to Sri Lanka, excluding tsunami aid, amounts to about $10 million annually.

Foreign Aid to Sri Lanka: CIDA's story

The story of CIDA's aid to Sri Lanka is an important one, however, because CIDA's assistance has been shaped, in part, by the conflict itself. The Canadian Government recognised the dismal conditions of human rights violations that characterised the country after the 1983 pogroms against Tamils in Colombo. By the late 1980s, human rights

---

[47] CIDA, 2001, pp. 3, 6.

[48] Joe Knockaert, "Canada making a difference in the world," CIDA presentation and overview, Victoria, Canada, 8 June 2004.

[49] Smillie, *op. cit.*, p. 15.

[50] Knockaert, *op. cit.* Even more interesting is this geopolitically charged headline, quoting the chair of the Canadian Senate Committee on National Security and Defence: "'We won't win unless aid money flows." The article rehearses arguments made by members of the Senate committee that Canada cannot track its aid in Afghanistan because it gives a lump sum to the Afghan government which decides how and where to spend it. Without the optics of Canadian aid going into the dangerous Kandahar region where Canadian soldiers are stationed and being killed on a regular basis, Senate members argue, Canada cannot win the hearts and minds of Afghans. In this instance, whether or not the government of Afghanistan is corrupt, unstable, or has protectionist economic policies does not appear to be relevant to aid disbursement. See Daniel LeBlanc, "We Won't Win" Unless Aid Money Flows," *The Globe and Mail*, 6 October 2006.

atrocities were atrociously high and Canada was receiving high numbers of refugee claims from Sri Lankan Tamils.[51]

The Canadian government was, in short, unwilling to provide bilateral aid directly to a recipient government that had committed so many of those violations.[52] The LTTE also played a part in human rights violations and the violence that characterised Sri Lanka at that time.

Canada is the only bilateral donor operating in Sri Lanka that directs its contribution solely through civil society organisations and non-governmental channels. While it does engage government ministries by offering expertise and training in specific areas, Canadian aid is an unusual configuration of bilateral assistance: no funds are channelled directly from government to government. CIDA's bilateral contribution to Sri Lanka is modest, at roughly CAD$5.5 million per year. This bilateral commitment is not nearly significant enough to put Canada on Sri Lanka's 'top ten' list of donors of net official development assistance, topped by Japan (US$167 million) and followed by the Asian Development Bank (US$94 million) and World Bank (US$59 million).[53]

Additional multilateral funds to the International Committee for the Red Cross (ICRC), the Canadian Red Cross, and *Médecins Sans Frontières* (MSF Holland) push total Canadian contributions in Sri

---

[51] Nira Wickramasinghe, "Sri Lanka: the politics of purity," Open Democracy Website, accessed 11 November 2008 at www.opendemocracy.net/democracy-protest/ sri-lanka_state_4105.jsp .

[52] UTHR(J) [University Teachers for Human Rights, Jaffna], "Report No. 1," (1989), accessed 8 July 2008 at http://www.uthr.org/Reports/Report1/Report1.htm. The Canadian position was outlined in Interviews #2 and #18 during 2002 in Colombo, Sri Lanka.

[53] D. Sriskandarajah, "The Migration-Development Nexus: Sri Lanka case study," Paper prepared for the Centre for Development Research Study: Migration-Development Links: Evidence and Policy Options Magdalen College, Oxford, United Kingdom, 2002.

Lanka up to over CAD$10 million.[54] Much of CIDA's focus being on the conflict-affected Northern and Eastern provinces where more Tamils are concentrated. As with most international donors, CIDA is careful, however, to operate in the Sinhala-dominated South as well. A senior manager for South Asia programming at CIDA headquarters in Canada, however, noted that it has the fourth largest bilateral aid program in Sri Lanka's Northeast.[55] Longstanding ties between the two Commonwealth countries and neutral political positioning between the Government of Sri Lanka and the rebel Tamil Tigers (LTTE) during the last two decades of war have been the basis of relative political neutrality until 2006 when Canada and the European Union joined Britain and the United States in banning the LTTE as a terrorist group. Domestic political considerations in Canada shape these aid disbursements through the large, well-organised, though by no means internally unified Sri Lankan Tamil 'asylum diaspora.'[56]

"Development success can be a peace-building failure," said one CIDA official in Colombo in 2002.[57] CIDA has avoided funding capital-intensive projects, the conventional 'bricks and mortar' approach, opting for 'knowledge-based' training and projects to enhance institutional and community-based capacity-building for more than a decade.[58] The latter have smaller budgets with budget line items that are much more difficult to divert in the context of conflict. Assurances that international assistance is delivered in a 'conflict sensitive' manner are key in this context.[59] Bilateral aid to the Government of Sri Lanka

---

[54] Interview #17, 2002, Colombo, Sri Lanka.

[55] Interview #13, 2002, Ottawa, Canada (interviews were technically held at CIDA's headquarters in Hull, Quebec, Canada).

[56] Christopher McDowell, *A Tamil Diaspora: Sri Lankan Migration, Settlement and Politics in Switzerland* (Providence, Rhode Island: Bergahn Books, 1996).

[57] Interview #2, 2002, Colombo, Sri Lanka.

[58] CIDA, *Program in Sri Lanka – An Overview*, 2nd ed., August 2003.

[59] Vance Culbert, "Civil Society Development vs. the Peace Dividend: International Aid in the Wanni," *Disasters* 29, no. 1 (2005): 38-57.

can be used to subsidise existing health, education or social services so that funds are redirected for defence purposes. Likewise, the LTTE informally 'taxes' all goods and services within its territory, including wages and materials for humanitarian and development work.[60] Projects with large capital budgets are at risk of being taxed to an even greater degree.

Making the Cut: The Case of Sri Lanka

During interviews with senior managers at CIDA headquarters in Canada during 2002, I was told that Sri Lanka would be cut from the Canadian aid map when CIDA reduced the number of countries that receive its aid in accordance with aid effectiveness policy.[61] According to three informants at CIDA's headquarters, Sri Lanka was 'developed enough' to be 'graduated' from Canadian development assistance.[62] A transition plan for Sri Lanka was underway to allow CIDA to exit gradually from the modest bilateral support it provides.

In Colombo during July 2002, the Canadian High Commissioner for Sri Lanka noted that aid effectiveness had not yet trickled down from headquarters in Ottawa to Colombo. She did not confirm that Canadian aid to Sri Lanka would taper off, but instead saw Canada's aid role as one of "constructive engagement" whereby Canadian aid had and would continue to reduce tensions in relation to the conflict.[63] Similar views were echoed by CIDA personnel at the Colombo office. Canadian bilateral aid, channelled largely through national non-governmental organisations, and its multilateral assistance, provided through players like *Médecins Sans Frontières*, would continue to be

---

[60] Kristian Stokke, "Building the Tamil Eelam State: Emerging State Institutions and Forms of Governance in LTTE-controlled areas in Sri Lanka," *Third World Quarterly* 27, no. 6 (2006): 1034.

[61] Interview #12, 2002, Ottawa, Canada.

[62] Interviews #12 and #13, 2002.

[63] Interview #18, 2002.

used for peace-building purposes in a geopolitically non-strategic location.

More directly, CIDA funded experts from the Forum of Federations, a global network on federalism based in Ottawa, to discuss constitutional options, comparative case studies, and prospects for a potentially federalist state.[64] This aid, coupled with assistance drafting an official languages act, marked a shift from no bilateral aid to high-level 'in kind' bilateral assistance, but still no hard cash for Sri Lanka government programming.[65] Very quietly, CIDA was trying to wrap up its bilateral programs in Sri Lanka and gradually exit the country altogether.

In 2002, messages about CIDA aid to Sri Lanka in Ottawa and in Colombo were very different, despite interviews being held within one month of each other. Yet, there seemed little question that aid effectiveness policy would eventually make its way south.[66] And then, in the early hours of Sunday, 26 December 2004, the tsunami hit.

"The tsunami changes everything," said the new Canadian High Commissioner for Sri Lanka in a 2005 interview.[67] Canada was not leaving Sri Lanka any time soon in terms of bilateral aid, technical assistance, and knowledge transfers. She noted that the post-tsunami visit of then Prime Minister Paul Martin in January 2005 cemented Canada's and CIDA's role for the next decade.

---

[64] Interview #18, 2002; R. Chattopadhyay, "A new peace initiative in Sri Lanka," Forum of Federations, (2002), accessed 8 July 2008 at www.forumfed.org/en/libdocs/Federations/V2N5-lk-Chattopadhyay.pdf. Constitutional expert, Dr. Peter Meekison, and former Ontario Premier, Bob Rae (now a Canadian MP), participated in several consultations, albeit these were held exclusively with the Sri Lankan Government.

[65] Interview #25, 2005.

[66] In 2005, I spoke with a CIDA staff member who had worked on various bilaterally funded projects in 2002. She told me that *projects* had been largely completed, with few new pots of money for project-based aid. Rather, a *programme-based* approach had been adopted, though the impact of this remained unclear to her.

[67] Interview #25, 2005, Colombo, Sri Lanka.

A CIDA vice-president in Ottawa echoed this point and noted that the tsunami was a factor in deciding to include Sri Lanka on CIDA's 'top 25' list "because of the global lens" and visibility of the crisis to Canadians. He also remarked that policy formulation was not wholly a rational decision and depended in part on the outcome of "political vetting" and support from the Prime Minister's Office. The Canadian Government's International Policy Statement which included CIDA's list of 25 development partners was published after the tsunami. The CIDA vice-president added that "one of the factors they looked at was diaspora." Sri Lanka stayed on the list, in part, because of the Tamil diaspora in Canada.[68] Nepal, another contender for the development partner slot allocated to Sri Lanka according to the same vice-president, was not selected because "there is no Nepali diaspora in Canada to speak of." At the time of the decision about which countries would be on CIDA's development partners list, Nepal was also in a state of political turmoil with no solution in sight. The CIDA informant added that Sri Lanka's apparent possible path to peace at that same moment helped it edge out Nepal, a much poorer country.

Transnational geopolitics and relative political stability at a key moment emerge as decisive factors for continued bilateral funding from CIDA. The intersection of two 'disasters', the tsunami and the conflict, changed CIDA's funding patterns in Sri Lanka. I now turn to the impact of the tsunami on foreign aid and on domestic politics more generally, beyond the specific case study of CIDA.

The Politics of Post-Tsunami Aid

Initially because of the CFA [Ceasefire Agreement] everyone seemed to forget about the conflict and worked together straight after the tsunami. But when certain issues came into

---

[68] Interview #35, 2005, Ottawa, Canada. Bastian, 2007 also discusses the constructive role of the diaspora on page 189; see also Kristian Stokke's analysis of the diaspora's role in "After the Tsunami: A Missed Opportunity for peace in Sri Lanka?," *NIASnytt* 2 (2005): 12-20, and Hyndman, 2007, *op. cit.*.

play, then it brought about the picture that tsunami aid was not conflict-sensitive.[69]

The tsunami actually helped in aggravating the situation.[70]

Respondents in the 2007 research project were asked if they thought the tsunami had any relation to the conflict in Sri Lanka, and if tsunami relief efforts played any role in a) fuelling conflict; b) mitigating conflict; and/or c) distracting people from the conflict. Evidence of all these scenarios in the Sri Lankan context has been witnessed in the pre-tsunami period, before the Ceasefire Agreement was signed in 2002.[71]

In the wake of the tsunami and in the context of renewed conflict, three main issues were identified by aid officials canvassed: 1) inequalities were either created or exacerbated in the wake of tsunami aid; 2) the buffer zone policy became a source of tension; and 3) the failure of P-TOMS was indicative of renewed conflict. Each of these is addressed in turn.

Exacerbating Inequalities: The Forgotten IDPs

That Internally Displaced Persons (IDPs) who were not tsunami-affected received little, if any, aid has been a source of tension across Sri Lanka. One of the donors interviewed for the IDRC project emphasised that his agency maintained a balanced program, ensuring that funding went out to all affected parts of the country.[72] Such geographical distributions across political lines – between areas controlled by the LTTE and those by Government forces – did not, however, translate into equitable distributions across the category of 'displaced.' With most humanitarian aid in 2005 earmarked for the tsunami-affected

---

[69] Interview #102, 2007, Colombo, Sri Lanka.

[70] Interview #105, 2007, Colombo, Sri Lanka.

[71] See Malathi de Alwis and Jennifer Hyndman, *Capacity-Building in Conflict Zones: A Feminist Analysis of Humanitarian Assistance in Sri Lanka* (Colombo: ICES, 2002).

[72] Interview #103.

population, war-displaced people in Puttalam, Mannar, and near Anuradhapura saw no change in their situation despite the huge opportunity that generous tsunami aid represented. Assessing the urgency of humanitarian needs is, of course, necessary for planning and providing assistance, but the humanitarian crisis of war and the unmet needs it created well before the tsunami were largely neglected afterwards.

At the same time as conflict-affected displaced people (not affected by the tsunami) were ignored, politically charged places of tsunami-affected people attracted disproportionate international attention and resources. The concentration of aid in constituencies represented by senior government ministers is not a focus of this analysis, but one aid official did name Hambantota on Sri Lanka's southern coast as a place that received far more than its share of aid: "In comparison to what the North and the East received, more funds went out to the South, especially Hambantota area."[73] In December 2005, less than a year after the tsunami, *The Sunday Times* [of London] reported that aid agencies were constructing 4,478 houses in Hambantota, even though only 2,445 were needed.[74] Hambantota is a political constituency represented by then Prime Minister, now President, Mahinda Rajapaksa. Meanwhile, in the predominantly Tamil district of Ampara, arguably the hardest hit by the tsunami, *The Sunday Times* said, only 3,136 houses are being built for the 18,800 families whose homes were destroyed. There is no question that politics distorted aid distributions.

Buffer ('Set Back') Zones

When the buffer zone made its entrance after the waves had left behind the destruction [in January 2005], it was known by another name, less popular—set back zone. It most certainly has lived up to that title. –Amantha Perera[75]

---

[73] Interview #105.

[74] Dean Nelson, "Old prejudices keep tsunami aid from Tamils," *The Sunday Times*, 18 December 2005, accessed 19 July 2008 at http://www.timesonline.co.uk/tol/news/world/article767365.ece.

[75] Amantha Perera, "The Buffer Zone Fiasco," *The Sunday Leader*, 25 December

In January 2005 the Sri Lankan Cabinet of Ministers legislated buffer zones, ostensibly as a public safety measure against the potential devastation of another tsunami.[76] In the South, dominated by a Sinhala majority and international tourism, a one hundred-metre buffer zone was established. In the Tamil- and Muslim-dominated Eastern Province, where tsunami-related devastation and damage proved greatest, a two hundred metre buffer zone was declared. In the LTTE-controlled Wanni, 400 metre set-backs were declared, though these were later amended to 300 metres, ostensibly an act of LTTE so

Figure 2: No buffer zones for tourist hotels, Unawatuna, Southern Sri Lanka

vereignty. In the South and the East, where population density was high and land scarcity acute, the buffer zones were highly contentious.[77]

2005.

[76] Centre for Policy Alternatives. "Landlessness and Land Rights in Post-Tsunami Sri Lanka," Commissioned by International Federation of the Red Cross (Colombo: 16 November 2005).

The buffer zones created contested political spaces characterised by polarised party politics, and an "opportunity to fish for votes."[78] The public safety rationale for the buffer zones seemed particularly thin. As the Institute for Policy Studies in Colombo stated, if public safety was the prevailing aim, the buffer zones should have been equivalent for all areas.[79] This apparent "fix" served instead to fan the flames of political controversy between the major political parties and among the various ethno-national groups that constitute the Sri Lankan populace, namely Sinhalese, Tamils, and Muslims.

So much more land was rendered unsuitable for housing in the Eastern Province compared to the Sinhala-dominated South that tsunami-displaced people in the East, minority Tamils and Muslims, claimed discrimination by the Sinhala-dominated government. In the South, tourist hotels and resorts not destroyed by the tsunami were allowed to rebuild within metres of the high tide water line (see figure 2), whereas residents could not rebuild their homes within 100 metres of the same line. In the East, narrow strips of land between the sea and lagoons, called littoral, or *eluvankaral* in Tamil, once home to separate Muslim and Tamil villages were declared "unliveable" by the government through the buffer zone legislation. The hinterland, or *paduvaankaral*, inland from (and west of) the lagoon is occupied largely by Sinhalese settlers who moved there through government colonisation schemes during the more nationalist periods of the 1950s and 1970s.[80]

---

[77] See Hyndman, 2007, *op. cit.* for a fuller analysis of buffer zones. The World Bank estimated damage in geographic proportions immediately after the tsunami: 40% along the East Coast, 30% along the South Coast, 20% in the North, and 10% along the Western Shore.

[78] Kingsley A. de Alwis, "The 100-metre rule – What's the logic?," *The Island,* 25 May 2005.

[79] Institute for Policy Studies, *Sri Lanka: State of the Economy 2005* (Colombo: Institute for Policy Studies, October 2005).

[80] D. B. S. Jeyeraj, "The State that Failed Its People," *The Sunday Leader,* 6 February 2005, p. 17.

Layers of political geography, from colonisation schemes to conflict-related displacement, preceded the tsunami which, in turn, uprooted and dispossessed so many more. But the buffer zones were an unexpected source of displacement. With few prospects of rebuilding, people found themselves displaced again, this time by a policy that only stirred tensions further. In February 2006, the newly renamed Reconstruction and Development Agency (RADA) under President Rajapaksa announced that the buffer zone 'set back standards' would be relaxed and retrofitted to follow the Coastal Zone Management Plan of 1997.[81] Nonetheless, considerable mistrust had been mustered by the buffer zone plan by that time.

P-TOMS: The 'Joint Mechanism' that couldn't

In one interview, an aid official responded, "there have been many missed opportunities since the tsunami. For example, the P-TOMS. There was no agreement between the Government of Sri Lanka (GoSL) and the LTTE, but this could have clearly been a breakthrough."[82] After negotiations between the Government and LTTE began to dissolve in 2003, P-TOMS was the most public and political set of negotiations to take place. Its aim was to create a 'joint mechanism' to determine the distribution of tsunami aid. The World Bank had assessed the cost of reconstruction and recovery at approximately $1.5 billion.

How to allocate these funds emerged as one of the biggest hurdles between the government and the LTTE. The joint mechanism was proposed at an international donor forum in May 2005 and was a memorandum of understanding (MoU) that set out terms for a working relationship between the GoSL and the LTTE. Representation from the Sri Lankan Government and the LTTE was augmented by that of Muslim political parties. Muslim communities living along the

---

[81] *Sunday Times*, "Coast Conservation Buffer Zone Limits Relaxed," a RADA ad in *The Sunday Times*, 5 February 2006.
[82] Interview #107.

devastated East coast were disproportionately hit by the tsunami, given their overrepresentation in that region.[83]

Signed in June 2005, the P-TOMS MoU quickly became a contentious agreement, galvanising Sinhalese nationalist sentiment in opposition to it because of the legitimacy it potentially conferred on the militant Tamil nationalists, the LTTE. The Sinhala nationalist JVP party, in a coalition with the sitting government, petitioned the Supreme Court in July 2005 on the grounds that P-TOMS violated the rights of Sri Lankan citizens and the territorial integrity of the state on the following grounds: 1) the LTTE was a terrorist organisation and not a governmental entity that could participate in such an agreement; 2) committees described in P-TOMS were governmental in nature and could not legally do the work they were charged to do without constitutional changes; 3) donor funds belonged to the country and could not be controlled by an outside agency like the World Bank [as was outlined in the MoU]; and finally, 4) the treatment of persons within the tsunami disaster zone (TDZ) discriminated against tsunami-affected persons outside the TDZ.[84]

The Supreme Court ruled in favour of the plaintiffs, and PTOMS died a quiet death with no shared mode of decision-making was ever introduced. As Bastian's book suggests, however, just three years earlier, negotiations between the Government and LTTE proceeded without such an outcry. At the 'Regaining Sri Lanka Strategy' donor meeting in Tokyo in 2003, $4.5 billion of foreign aid was pledged for reconstruction. No one protested the fact that foreign assistance would fund the Strategy. And 80% of Sri Lankans polled supported negotiations at the time.[85]

---

[83] Jennifer Hyndman, "Siting Conflict and Peace in Post-tsunami Sri Lanka and Aceh, Indonesia," *Norwegian Journal of Geography* (forthcoming 2009).

[84] Margo Kleinfeld, "Misreading the post-tsunami political landscape in Sri Lanka: The myth of humanitarian space," *Space and Polity* 11, no. 2 (2007): 179.

[85] Bastian, 2007, *op. cit*, p. 156.

These three sources of tension and conflict identified by the majority of aid officials polled provide a context for the more specific story that follows. Clearly, the longstanding conflict in Sri Lanka is inseparable from the tsunami and humanitarian aid that followed it. The IDPs not affected by the tsunami, some displaced by fighting for upwards of 18 years, were largely ignored in the post-tsunami period. The buffer zone policy, supposedly to ensure public safety, only heightened mistrust toward the Government. And the promise of P-TOMS was quashed by the acrimonious political sentiment of the day, when the Ceasefire Agreement had all but collapsed.

Conclusion

One of few positive outcomes of the 2004 tsunami has been its influence on Canada to continue bilateral aid, however modest, to Sri Lanka. The tsunami, in conjunction with the geopolitical influence of diaspora, the potential for peace, and the visibility of Canada's role in responding to the disaster, could be characterised as a crisis of exceptionalism. All of these factors shifted Canada's position on bilateral aid just enough to include Sri Lanka [and Indonesia where Aceh Province was very hard hit by the tsunami] on its list of 25 priority development partners. One lesson of this story is that aid decisions are based, in part, on policies such as aid effectiveness but are also made 'on the fly' based on assessments of rapidly changing conditions of disaster and related geopolitical factors such as the significant size of the influential Sri Lanka Tamil diaspora and the strong support of the Canadian public for those affected by the tsunami, "white death" or otherwise.[86]

While CIDA is not an exemplar in terms of its precise response in post-tsunami Sri Lanka, it is certainly part of the aid effectiveness alliance of donor countries, including Britain, the Netherlands, and the

---

[86] Alison Mountz's ethnography of the Canadian state, specifically the Department of Citizenship and Immigration, traces the formation of 'policy on the fly' in more detail; see A. Mountz, "Embodying the Nation-State: Canada's Response to Human Smuggling," *Political Geography* 23, no. 3 (2004): 323-345.

United States. Its decision to break with the principles of aid effectiveness and its own policies it adopted in the early 2000s suggests that practical geopolitical considerations on the ground can trump neoliberal ideals that circulate among donors in the Global North.

The analysis in the last part of this paper, in which the relation of tsunami aid to tensions and conflict is probed, calls up other 'lessons learned'. Where war precedes a so-called 'natural' disaster, the casualties of the first humanitarian crisis cannot be ignored in the wake of the second. The IDPs in Puttalam, many of whom fled their homes after LTTE attacks in 1990, still receive government rations in their temporary homes of the Western Province.[87] They are the forgotten IDPs, the 'old caseload' as it were. Humanitarian aid may be earmarked for specific crises, but every effort must be made to ensure sufficient flexibility in the terms of its provision so as not to increase inequalities, tensions, or resentment among those displaced by humanitarian crises. Without a strong understanding of "new political formations emerging on the global periphery," humanitarian aid will be incorporated into the fabric of political violence.[88]

Buffer zone policies that appear to enhance public safety will be perceived as an opportunistic land grab if they are not consistent across regions and rationalised by scientific means. In Sri Lanka, the buffer zone edicts were a disaster in themselves, displacing people yet again by preventing them from rebuilding homes next to the shoreline. And P-TOMS, the joint mechanism, was a barometer of the negative political climate in which it was negotiated, not a harmful policy or practice per se.

Foreign aid to Sri Lanka will remain a vital source of debt relief, development assistance, and humanitarian relief in the coming years. In 2006, total foreign aid disbursements to Sri Lanka, a country of less

---

[87] Cathrine Brun, "Finding a place. Local integration and protracted displacement in Sri Lanka," Doctoral thesis, NTNU, Trondheim Norway, 2003.

[88] Mark Duffield, "The Symphony of the Damned: Racial Discourse, Complex Political Emergencies and Humanitarian Aid," *Disasters* 20, no. 3 (1996): 173.

than 20 million people, amounted to over $1 billion.[89] The stakes are high. Probing the global discourse of foreign assistance, such as aid effectiveness, alongside humanitarian crises including the tsunami and conflict in Sri Lanka, allow us to see exactly what trickles down, what does not, and how geopolitics influences the provision of aid at the end of the day.

Jennifer Hyndman is Associate Professor of Geography at Simon Fraser University in Vancouver, Canada.

---

[89] See tables in Bastian, 2007, *op. cit.*, especially p. 193.

'Communities' West and East: Post-Tsunami Development Aid in Sri Lanka's Deep South East[1]

Pradeep Jeganathan

Abstract
'Community' and 'the consulted community' have become naturalized in aid discourses and practice produced by INGOs and
their local partners and are intertwined with other naturalized
categories like empowerment and participation. While I am in
agreement with critics, such as Stirrat, who have traced the
Christian and Orientalist origins of these categories, I argue
that the colonial and nationalist history of these constructions
has been neglected by these critics. I outline a coloni-
al/nationalist genealogy of these categories, and link it to the
self–presentation of 'community' in relation to post-tsunami
aid. In the three contemporary examples considered in Sri
Lanka's deep South East, I find that these practices of making
'community' are also practices of the hierarchical reproduction
of social order and the making of an authoritative 'voice', which

---

[1] I am most grateful to the editors of this volume, Malathi de Alwis and Eva-Lotta
Hedman, for their questions and suggestions as well the other researchers on this
project for their comments on an earlier version of this paper at the mid-project
workshop in Colombo. I am also grateful to Malathi de Alwis for discussions of my
arguments and assistance with sources. Invaluable field assistance was provided by Research Associates Vidharshana Kannangara, Prabath Kumara and Athula Samarakoon.
Generous research support was provided by IDRC. Following anthropological con-
vention, several identifying place names have been changed; they will however be
easily located by specialists who are familiar with the literature cited. My title is a play
on Henry Maine's classic, *Village Communities East and West* (London: H. Holt &
Co, 1870).

is predicated on doxic silencing and repression of other voices and groups.

Introduction

In the last few decades, there has been an increase in 'development aid' channeled directly by international donors in support of infrastructural construction such as roads, bridges or harbors or more explicitly humanitarian projects such as schools, single family dwellings, or micro entrepreneurial activity from soap making to goat rearing, all over Sri Lanka. The devastation and destruction of the post-tsunami moment and absence of sustained hostility on the part of the Government of Sri Lanka to international aid efforts, of whatever variety, allowed for such aid to increase to yet another level. These efforts, in their discourse and practice, often invoke a particular conception of 'community' which is taken to mean small-scale rural settlements which are beneficiaries of such aid. Furthermore, 'consultation' with 'communities' is often posited as a crucial part of the planning and execution of these aid projects.

This paper attempts to formulate a critical analytical framework within which this category of '(village) community' can be re-described and re-understood. The this critical work is both theoretical and concrete in nature and the arguments draw on empirical research carried out in three somewhat interlinked settlements in Sri Lanka's deep south and south east that were affected by the tsunami, and then were the intended recipients of aid. By offering what I call a 'critical analytical framework' that may lead to such a 're-description' of 'community,' I attempt to address a serious error in the discourse and practice in the larger project of 'development aid' produced in general by International Non Governmental Organisations (INGOs) and International Government agencies.

The critical framework I offer goes beyond the kind of 'assessment' that is usually offered by an 'outside evaluator,' a common subject position in aid discourses. It is my suggestion that many of these evaluations are insufficiently robust to grasp the reality of the social formations they are working with. In fact, the discourses that govern the received understanding of aid programs in relation to small scale settlements have together with the commissioned literature of 'evaluation,' become a language of social description, in and of itself, self-contained and bounded as it were, not rooted in any well known tradition of critical social science. Roderick Stirrat points to this when he writes "… at times it appears that there are two separate literatures on rural Sri Lanka, one produced by academics and another by consultants. They rarely refer to the work of each other."[2] There is then, an extraordinary need for fresh critical analysis that can inform aid discourse and practice. In fact, Stirrat and his colleagues have in recent years made valuable contributions to the critique of categories like "participation," "empowerment," and "community" and I have benefited from their argument that these categories can be linked genealogically to those of missionary Christianity or orientalist or colonial knowledges.[3] There is however, a gap in Stirrat's arguments between the discursive origin of categories like 'participation' and 'community' in Christian or orientalist discourse and their actualization in practice, before the advent of INGO led developmental projects. In the schema of Stirrat and his colleagues, these categories are constructed at one chronological point, well over a hundred years ago, and then actualized in developmental practices at the current time and the very recent past. What is missing is an account of the non-discursive practices that are simultaneous with the discourses of Christianity and colonialism, as the projects, in both senses, articulate with places like Sri Lanka. Or put another way, categories that have a discernible discursive life also have a

---

[2] Roderick Stirrat, "The Old Orthodoxy and New Truths," *Assessing Participation: A Debate from South Asia*, eds., Sunil and Nicola Bastian (New Delhi: Konark, 1996), p. 88.

[3] Stirrat, *op cit*. See also Heiko Henkel and Roderick Stirrat, "Participation as Spiritual Duty; Empowerment as Secular Subjection," in *Participation: The New Tyranny*, eds., Bill Cooke and Uma Kothari (London: Zed Books, 2001) and Malathi de Alwis, this issue.

non-discursive life, if the enunciation of these discourses are part of authoritative projects like missionary Christianity and colonialism. That non-discursive life has a history.

In this paper, I attempt to address this lacuna in relation to the actualization of the category 'community,' rather than point only to its discursive origins. My sketch of administrative deployment and nationalist appropriation will not be as comprehensive and rich as it could be but nevertheless, I suggest that the marking of some of the signposts in socio-historical construction of 'community' will strengthen our understanding of its current deployment. Ultimately, the aim of the socio-historical detour is to inform a history of the present in an attempt to understand the incompleteness of the category 'community' in current aid discourse and practice.

"Community" Reconsidered

What are the assumptions embedded in the category 'community' as it is deployed in current aid discourses? My point of departure in the unpacking I propose is the link between the concept of 'community' in the discourses under analysis and 'consultation.' Take, for example, this passage from an August 2006 USAID newsletter for a new, US$ 10.3 million bridge linking settlements in the deep south east -- that were studied during the course of the project -- with the rest of the country. The bridge was washed away by the tsunami but a serviceable, pontoon bridge was built back very quickly by Indian army engineers. Nevertheless, USAID spent considerable resources rebuilding what is a massive bridge in relative terms. Progress was described in this way: "Plans for the bridge were developed in consultation with the *communities*; women and all ethnic and religious groups were represented."[4] We see here a triad between community, representation and consultation. In another example, World Vision, a Christian

---

[4]"USAID: Tsunami Reconstruction Update Aug 2006," accessed in December 2006 at http://www.reliefweb.int/rw/rwb.nsf/ db900sid/YAOI-6UN4MS/OpenDocument, my emphasis.

INGO that worked in one of the communities studied, describes a set of organizational mechanisms they put in place to facilitate "community consultations." The INGO notes in a published report: "...World Vision formed a special team called the Humanitarian Accountability Team (HAT), which focused solely on community engagement and interaction."[5] Over 600 such "community consultations" were held by HAT teams, the report adds. This discourse is pervasive, it seems, and is taken up by other organizations that are not INGOs, in an imitative fashion. Colliers International, a global real estate company, also assimilates this discourse of consultation and community to describe its own post-tsunami philanthropic efforts in one of the settlements studied. A news report of a press conference held by Colliers, which summarizes and is written in the voice of project managers speaking at a press conference, emphasizes that "constant interaction, community participation ... at all stages of this project resulted in its success."[6]

There is a relationship, here, between 'community' and 'consultation,' that has to be de-naturalized if a critical framework of re-evaluation is to be constructed. Consider the logic of the matter: A community is a group that has something in common, a collective practice or collective representation. For example, in a small rural settlement, this may be a water source, such as a well or reservoir, that is part of both common practices and representations. How would we configure 'voices' of such a community through well known traditions of critical social science? "Ideology" (Marx), "hegemony" (Gramsci), "subjugated knowledge" (Foucault), "doxa" (Bourdieu), "repression" (Freud), would all be well known, well debated, analytical stands that would point to the unavailability of transparent "voice." It is not only that one community can have many voices, it is also that many voices may be unarticulated. But, in the discourses of community consultation that dominates descriptions of INGO Aid practice, the 'communi-

---

[5] World Vision Sri Lanka Tsunami Response, "Final Report: December 2004 – December 2007," p. 16

[6] *The Island*, 9 March 2007.

ty' and the 'consulted community' appear to be synonymous. In fact, there is an implicit sense that if enough effort is made through some kind of organizational mechanism, whether it be HAT teams or notice boards, that 'voice' will be universalized through the community. The model for this idea of community seems to be an American ideal, the town meeting – where all speak, all are heard, and disagreements are talked through. I suggest this model is naive in its understanding of social formations, be it American or Sri Lankan, 'Western' or 'Eastern'.

Colonial Constructions of Community

Nevertheless, such naiveté specifically in relation to 'community' does have a deep genealogy in authoritative discourses on South Asia. Sketching this lineage while attempting to understand its actualization in socio-historical processes is as I have said, the aim of this article.

As we embark on such a venture, it is important to understand another naturalization; the peasantry as such, or the inhabitants of a rural hinterland, need not be thought of as 'dispersed' communities, whether organized around 'villages' or not. As Bernard Cohn has shown in his classic essay on the subject,[7] the first layer of formal, serious, colonial knowledge on India took caste, not community, as its major organizing category. The 'village community' as a keystone of the colonial social construction of south Asia, came slightly later, but was in any event, a quite separate strand of knowledge from the caste-based social construction.[8] An important moment of the rising into authoritative prominence of the 'village community' view was the publication

---

[7] Bernard Cohn, "Notes on the History of the Study of Indian Society and Culture," in *An Anthropologist Among the Historians and Other Essays* (Delhi: Oxford University Press, 1987): pp. 136-171.

[8] The distinguished anthropologists, Louis Dumont, who in an effort to construct an alternative theory of Indian caste, clears away much of the colonial literature on 'village community', which Cohn draws on in his work. See Louis Dumont, "The 'Village Community' from Munro to Maine," in *Contributions to Indian Sociology* 9 (December 1966): 67-89.

of Henry Sumner Maine's *Village Communities of East and West*,[9] in 1870, which in turn drew on earlier colonial accounts, such as Metcalf's.[10] Maine's project, on the one hand, followed in the epistemological path carved out by other European and colonial scholars through the middle of the nineteenth century: it was to locate what was observed in contemporary India in the past of Europe. India was thought to be Aryan and the Indian village community was thought to be a survival of what had once existed in Europe -- a little, self contained, self governing, 'republic' that had lasted through time, surviving invasion after invasion, the veritable soul of the country side. While the career of this idea, and debates over it in India, need not detain us unduly here, the importance of Henry Maine's construction of the 'village community' for Sri Lankan administrative, political and nationalist practice, should be underlined.

As Vijaya Samaraweera has demonstrated in his detailed reading of the matter,[11] the Village Communities Ordinance of 1871, a piece of legislation enacted in the last administrative years of Governor Hercules Robinson, owes a great deal to Maine. The motivation of the legislation was to put an end to what was seen as excessive litigation on matters of land and other disputes in the countryside. Maine's idea of a 'village community' as a self governing, autonomous entity was seized upon by colonial officials in Ceylon who argued that by creating 'village councils' (tribunals) where minor disputes could be adjudicated, litigation in the courts would be reduced and the 'village' thought to be, as I have suggested before, an Aryan survival from ancient times, could be revived and stabilized. This was in spite of clear differences between the Sri Lankan case and Maine's own flawed construction of the Indian case —village panchayets were caste based in a way that never ex-

---

[9] Maine, *op cit*.

[10] Cohn, *op.cit.*, p. 159.

[11] Vijaya Samaraweera, "Litigation, Henry Maine's Writings and Ceylon Village Communities Ordinance of 1871," in *Senerath Paranavithana Commemoration Volume*, eds., L. Premathilaka, K. Indrapala, J. E. Louhizen-deLeeuw (Leiden, 1978).

isted in Sri Lanka and land was not owned in common, one of Main's central assumptions. The ordinance was not welcomed by some articulate sections of the Sri Lankan intelligentsia at the time; Samaraweera reports widespread protests, which were however muted and civil, on several aspects of the legislation. An important objection was to the colonial proposal to appoint a new village council which would include a headman and other officials. Eventually, the headman, irrigation headmen and others to the council were appointed by the government. These appointees were local 'notables' who were picked by or approved by the Government Agent, an elite civil servant, who was usually British for most of the nineteenth and early twentieth centuries. Samaraweera has also pointed out that while certain kinds of local tribunals functioning under the overlordship of the nobility and ultimately the king had fallen into abeyance in 1870 in the Kandyan regions, they may never have existed in the littoral.

There are several points that are worth taking away from this complex colonial intervention. First, as a preamble, let us note that the protests and arguments over who would have the authority to adjudicate local disputes is surely part of the already apparent problem of rule that the ordinance of 1871 is trying to address; at the end of nineteenth century, local authority was not a settled matter. It may well have been under the sign of the kings of Kandy, but that royal authority always visible, significant and arbitrary, is replaced by increasing rule-bound administrative practice from London, via Colombo. Thus, there was a hollowing out[12] of the power of nobility from whom earlier local authority would have flowed. In fact, the whole colonial legal system, established from 1818 onwards, usurps the authority of even the previously 'ad-hoc and arbitrary' village councils in explicit ways, certainly with the establishment of minor courts in 1843. Fascinatingly, the 1871 legislation was a regressive one and clearly whatever authority had resided in particular local figures had now been disputed in the co-

---

[12] The idea is from Nicholas Dirks, *The Hollow Crown* (Cambridge: Cambridge University Press, 1987).

lonial courts, for several decades. The legislating and forming of village councils, sought to re-construct rural authority by a series of appointments, not only in the judicial realm but also in the moral and administrative realms. Colonial administrators were explicit that 'village councils' were to re-make moral 'community' through 'communal self government' within which crime would lessen and education would thrive. And they were continually guided by the transposition of Maine's ideas to Ceylon which implied that Sinhala Villages were as English Villages once were and needed to be guided back, in a 'patriarchal system' led by the 'natural leaders of the community'.[13]

This is not to say that the authority of village councils become settled and undisputed; it is to say that the 'village council' and the officials it spawns as minor administrators -- the village headman (*gamarala* or *arachchi*) and irrigation headmen (*vel vidane*) -- become the focal point of a history of local disputes. And so, 'village community' is born. This is yet another example of a rich series of examples of colonial constructions that makeup the socio-historical landscape of South Asia. The office of *vel vidane* was abolished by the post-colonial legislation of the Paddy Lands Act of 1958, and the village headman has now metamorphized into the *Grama Niladhari* (through *Grama Seveka*). But the watershed for the genealogy of the role and its context is 1871. Even a quick glance at the ethnographic literature demonstrates the history of disputed authority that flows through these petty administrative positions to the current day, a point I will return to.

Nationalism and Community

It would be a mistake however, to confine socio-historical processes that make 'community' in the wake of the colonial image of the Indo-Aryan village only to administrative history. My next step is to sketch out other nationalist strands of the social processes that actualize 'village community.'

---

[13] Government Agent Central Province, in 1871, quoted in Samaraweera, *op.cit.*, p. 198.

Two renowned and parallel figures are crucial for such an account: Anagarika Dharmapala and Ananda Coomaraswamy. Dharmapala, who straddled the monkhood and the laity in a unique, 'homeless' role which he carved out for himself, was both a Sinhala supremacist and an anti-imperialist in his discourses and practice, and was enormously authoritative among an increasingly politicized Sinhala counter public, in the late nineteenth century.[14] Coomaraswamy, an intellectual, geologist and art critic, was also, in my view at least, a cosmopolitan Sinhala nationalist of a very different ilk from Dharmapala.[15] Influenced by William Morris and the Arts and Crafts Movement as well as a wide swath of Orientalist literature, together with his own original investigations into the Sinhala craft traditions which he detailed in his *Medieval Sinhalese Art*, he provided both an intellectual and practical template from within which a cultural response to the colonial project could be forged.[16]

It is striking that both these figures, different as they are, seem to be influenced by Colonial images of the Indo Aryan 'village community.' Coomaraswamy's draws deeply from the well-known orientalist Sir John Phear's *The Aryan Village in Bengal and Ceylon*.[17] Phear's section on Ceylon is rather thin, reflecting the mere two years he says

---

[14] For a discussion of his Sinhala supremacy and nationalism, see Kumari Jayawardena's "Some Aspects of Class and Ethnic Consciousness in Sri Lanka in the late 19th and early 20th centuries," in *Ethnicity and Social Change in Sri Lanka*, ed., SSA (Colombo: Social Scientists' Association, 1985) and *The Rise of the Labor Movement in Ceylon* (Durham: Duke University Press, 1972). For a discussion of his reformist project, see Richard Gombrich and Gananath Obeyesekere, *Buddhism Transformed: Religious Change in Sri Lanka* (Princeton: Princeton University Press, 1988).

[15] Pradeep Jeganathan, "Disco-*very*: Anthropology, Nationalist Thought, Thamotharampillai Shanaathanan, and an Uncertain Descent into the Ordinary," in *Violence*, ed., Neil L. Whitehead (Santa Fe & Oxford: School of American Research Press, 2004). (Analytically, for me, a Sinhala Nationalist, need not be Sinhala; Coomaraswamy's father was Tamil, his mother English).

[16] Ananda Coomaraswamy, *Medieval Sinhalese Art* (New York: Pantheon Books, [1907] 1956)

[17] *Ibid.*, p. 23, for example.

he spent in the country and his picture of 'community' officials being 'chosen by the shareholders'[18] seems rather inaccurate and naïve. But, this tangentially informs Coomaraswamy's argument that village communities are ready for self-government. He compares a village to a parish council in England, and finds the latter wanting in forging 'communal interests' unlike the Sinhala example, where 'villagers work together in complex arraignments' and so are, in that sense, 'capable of self government.'[19] It is worth noting, as an aside, that both Phear and Coomaraswamy correctly jettison Maine's construction of 'village republics' as property holders in common, which informs to some extent the 1871 ordinance, but it does not stop them from constructing a romantic notion of 'village community'.

If Coomaraswamy's effort was to revive, Dharmapala's was to both revive and reform. A lot hangs on this distinction for reform for Dharmapala was a 'Protestant' project, that carried with it all the moral weight, and 'this worldly' ethic that would be associated with the term.[20] His writing, a compendium of which was a best seller in Sinhala going into dozens of printings in a few years, lays out in characteristic point form, a manifesto for village revival. Pivotal is a 'Village Protection Society,' which he insists should be founded in each village; this society is to be the vector of revival and reform. It should be led by the village monk, its lay members should lead a high moral life, it should educate both girls and boys and teach them a trade, it should ensure cleanliness is maintained everywhere in the village and in every sense. While Dharmapala's activism was focused elsewhere, it is clear that his manifesto did not go unheeded. H. L. Seneviratne, in his valuable and original account of the social actualization of this vision credits Wilmot Perera, a scion of a wealthy, reformist family, with

---

[18] John Phear, *The Aryan Village in Bengal and Ceylon* (London, 1880): 190.

[19] Coomaraswamy, *op.cit.*, 30

[20] Gombrich and Obeyesekere, *op.cit.*

inaugurating the first 'rural development society,' in Horana, in 1932.[21] Perera, in Seneviratne's careful estimation, is influenced by both Coomaraswamy and Dharmapala, with perhaps Coomaraswamy's *Medieval Sinhalese Art* being particularly important to him. But in the wake of Wilmot Perera's writings and work, come a series of monks associated with the Vidyodaya teaching monastery who begin far reaching activist projects resulting in the founding of a large number of associations and societies fully embedded in the project of rural 'upliftment' and 'development'. For example, one central activist monk held offices "of some kind in 275 societies, between 1917 and 1970."[22] Another "delivered ten sermons a week (on rural development) at 65 regular preaching sites, traveling in a car donated by (one of Wilmot Perera's relatives), Mrs. Jeramias Dias."[23] The devastation wrought by the malaria epidemic of the 1930s was an important catalyst for such work and the monks in their activist work could stress how morality, self-discipline and cleanliness, keystones in Dharmapala's exhortations, were crucial for eradicating the deadly disease.

It is important to stress that I am not attempting to argue that all this enormous work which results in active 'rural development' or 'village protection' societies throughout the country makes for uniform, harmonious 'revival' or regeneration; that would be romantic and naïve. My point is in fact the opposite, as is Seneviratne's. While some monkly activists understood the frameworks of political conflict and argument within which their work was situated,[24] and some did not, the central point I wish to make is that 'community' activity centered on moral 'good,' 'upliftment' and 'development,' has, throughout the last century, a variety of local, contested histories. It is not that the co-

---

[21] H. L. Seneviratne, *The Work of Kings: The New Buddhism in Sri Lanka* (Chicago: University of Chicago Press, 1999): pp. 56-129 in general, and on Wilmot Perera, p. 61.

[22] *Ibid.*, 67.

[23] *Ibid.*, p. 87.

[24] *Ibid.*, p. 101, nt. 32.

lonial idea of village community remains an ideological palimpsest on socio-historical life, it is actively engaged with, appropriated and indigenized, in ongoing micro political arguments on 'community'.

It is instructive to read how a careful ethnographer encounters these activist societies in rural Ceylon, about two decades after they had taken off but well before the ideas and practices of I/NGOs had entered the field. In Nur Yalman's preliminary description of a village in the Central Province where he did field work in the mid 1950s, which he has called Terunne, he tells us that there is both a Village Committee and a Rural Development Society that have interests in the settlement.[25] This is a fascinating instantiation of two 'constructions' of the 'village community' arising out of one social formation – one administrative and one subaltern nationalist – as I have sketched out above. The village committee then is the local manifestation of the *Gam Sabawa*, a council that comes into being after the 1871 Ordinance. Some of its leaders were appointed by the Government Agent's assistant, the District Revenue Officer (DRO), others were 'elected,' with the approval of the DRO. Nevertheless, it turns out that in Terunne, the Village Headman and the Chairman of the Village Council are the same person while his brother-in-law is the Irrigation Headman (*vel vidane*). They were wealthy and powerful in relation to the locality and managed to divert part of the government development aid channeled through the Committee, into their own coffers. All the posts on the Committee, Yalman tells us, 'were greatly coveted' for this reason. Clearly then, representation on the Council itself is part of a socio-historically situated claim about hierarchy, suitability and kin group. The representation of community is aligned with local power.

The Rural Development Society which might well be one founded by the kind of activist monks I have referred to above or a lay person influenced by the Vidyodaya rural development project's practices of reform and revival, works in a slightly different way. It too, in

---

[25] Nur Yalman, *Under the Bo Tree: Studies in Caste, Kinship, and Marriage in the Interior of Ceylon* (Berkeley: University of California Press, 1971).

this postcolonial moment, secures state aid which it expends on good works such as building houses, schools and latrines. There is an unmistakable stamp of Dharmapala's vision in this promotion of hygiene but it is not universally appreciated. Undoubtedly, the effort is an initiative of the petty bourgeoisie of one 'community' which is extraneous to the values of others. The villagers, Yalman writes, "happily accepted money for the latrines which they dug themselves, and then used for storing grain."[26] But, another 'community,' belonging to a lower ranked caste, was unsuccessful in securing resources from either the Village Committee, or the Rural Development Society. Again, we see the complexities of 'community' that is formed by socio-historical processes and contemporary political arguments which continues into the 'village reawakening movement' or *Gam Udawa* by the latter day reformer, the late President, Ranasinha Premadasa. This intervention, which is correctly located by Seneviratne[27] as both explicitly and implicitly linked to the work of a Vidyodaya monk, the Rev. Kalukodeyawe Pangasekere, focuses on those very 'communities' that would have been excluded from the first wave of activism: lower ranked caste groups.

This is the context within which the increasing magnitude of INGO-directed aid that followed the civil war and then again increased in magnitude after the tsunami, should be understood. My point in relation to the category of 'community' is not merely that a romantic, naïve construction of it prevails in INGO discourse and practice. It is also that representing 'community' as collective groupings in the form of associations or societies in order to obtain 'aid' for 'self-improvement' and 'moral reform' intertwined with graft and pilferage is a set of familiar practices in rural Sri Lanka, as is exemplified in the ethnographic literature.[28] The 'communities' that INGO aid seeks to

---

[26] Yalman, *op. cit.,*: 32

[27] Seneviratne, *op.cit.*, 79, nt.18.

[28] See for differing accounts, Michael Woost, "Developing a Nation of Villages: Rural Community as State Formation in Sri Lanka," in *Critique of Anthropology* 14, no. 1 (1994): 77-95, Tamara Gunesekere, *Hierarchy and Egalitarianism: Caste, Class and*

'empower' are already embedded in a meta-practice that is adroit in working with this construction and each of those 'communities' are, in turn, part of a complex micro-political process which cannot be transparently accessed through 'consultation.' Nor should we limit the field of this observation to the Buddhist majority areas of Sri Lanka, even though Seneviratne's account, for example, is by its argument, so limited. Rural Development Societies, Welfare Associations and Hindu temple and Mosque Management Committees are common in the north and east of Sri Lanka; they too are organized in very similar ways by the 'meta-practice' of representing community to themselves as well as outside donors, be it the state or philanthropists. It is of course a lack in the historical literature that prevents me from accounting in detail for their history which is however visible to the anthropological eye.[29]

Post–Tsunami Aid and Contemporary Community

I will now focus on three communities, two very proximate to each other at the tip of the Eastern Province of Sri Lanka, in the deep south-east of the island, and another at the extreme edge of the Southern Province, that is somewhat proximate to the other two. All three were differentially affected by the tsunami and also by the ongoing civil conflict. Seaville, a small fishing village, is also known internationally as one of finest surf beaches in the world and is visited by several hundred surfers at a time, during the season, which runs from March to September each year. Its Tamil population was decimated by the first wave of anti-militant state repression in the 1980s, but it became an uneasy zone of relative peace from the late 1990s, through the 2000s, with largely Muslim and Euro-Australian owned, small scale guest houses

---

Power in Sinhalese Peasant Society (London: Berg, 1994) and James Brow, *Demons and Development: The Struggle for Community in a Sri Lankan Village* (Tucson: University of Arizona Press, 1996).

[29] I draw here on Malathi de Alwis' extensive field work in the region, for more than a decade, and thank her for her insights. On the prevalence of various forms of associations in the Eastern Province, see Denis Macgilvray's *Crucible of Conflict* (Durham: Duke University Press, 2008).

and restaurants catering to surfers, holding sway against the war. Greenwood, another settlement 20 kilometres south of Seaville, has resisted tourist-friendly capitalization but has a fascinating, mixed Sinhala-Tamil community which has deep historical roots and a classic and venerable tradition of worship. This mixture has been respected as an oasis by both Tamil militants and the Sri Lankan armed forces, throughout the conflict. This makes both Seaville and Greenwood, in different ways, important counter-examples in a war zone. Seaville was devastated by the tsunami, but Greenwood was less affected. Yet, both have received considerable outside aid, given the surfeit of available aid and also the strong presence of a Sinhala community in Greenwood. Greenwood's economy is centered on agriculture, with paddy farming dominating. Fishing is a small component as well as state security services, given the Sinhala presence in the community coupled with its proximity to zones of conflict. Fishtown, another small settlement in the deep south of the country at the extreme edge of the Southern Province, is separated from Greenwood by a natural reserve. It, like Seaville, is a Muslim majority community (but in this instance, Malay and not Moor), but unlike Seaville, is dominated by a Buddhist temple that has considerable symbolic authority; though only built in 1956, this temple claims a link to the golden age of "Sinhala" history through a strategic symbolic association. It is thus an important auxiliary stop for Buddhist pilgrims on a southern tour of important religious shrines. These pilgrims, and also a small number of affluent, secular, nature lovers, who, given the close proximity of the wild life reserve to the town, may patronize its up market hotels and guest houses and contribute to the local economy. But Fishtown's mainstay is fishing, given the minor harbor that was first developed there, in 1983. Fishtown was hard hit by the tsunami with over 50 houses on the beach front being wiped out. It has been the site of several aid projects, including one which is internationally renowned.

The Fishtown Harbor

In what appears to be a press release, published in Sri Lankan newspapers as a news item, in late January 2008, USAID marks the opening of a new mechanics workshop for the Fishtown harbor, which is, it is said, 'a new beginning for the fishermen' of Fishtown.[30] Recalled also, is another project completed two years before, the dredging of the harbor, which the Mission Director of USAID says forged strong relationships with the 'people' of Fishtown which allowed the work on the mechanics workshop to proceed so well.[31] The 'collaboration' of the Ceylon Fisheries Harbor Corporation (CFHC) and the (Fishtown) Management Committee was complimented. "[W]e are optimistic," the Director concludes, "that the relationship between the CFHC and the *community* led committee will continue to blossom and evolve, making the habour a true gathering place for all members of the *community*."[32] However, a year later, another press item, which does not appear to be an agency handout tells a sorry story of the Fishtown Harbor, which is "not more than a swimming pool. The repair workshop, the ice plant and the fuel storage ceased to function ..."[33] This report is diametrically opposed to the glowing picture painted in the USAID press release. How could this be?

Since field work had ceased in Fishtown, about three months before this report was published, it was not possible to check its exact accuracy. Nevertheless, research done over the previous year can shed some light on the dramatic difference of the two reports.

The Fishtown fishing habour was initially developed by the Japan International Corporation Agency (JICA), in 1982-1983. It was opposed by the chief Buddhist monk of the Fishtown temple, but it is

---

[30] The 'news' item appears identically in both in the *Ceylon Daily News* of 25 January 2008 and *The Island* (Financial Review) of 29 January 2008, implying that it was a press release, published with no editing, in the press.

[31] *Ibid.*

[32] *Ibid*, my emphasis.

[33] *Daily Mirror*, 9 January 2009.

said that the Japanese finessed this problem by donating a building to the temple. Nevertheless, apparently because of a serious design flaw that did not take into account the movement of tides and seasonal wind patterns, the harbor began to silt up. A rehabilitation project, which included two extensions of the breakwater and a dredging of the harbor, took place between 1991-1994, also funded by JICA. Altogether, a sum of over ¥1.2 billion (approximately US$ 100 million) was spent on this work, according to JICA.[34] Nevertheless, the harbor continued to silt up. For the harbor to be used, it needs to be continuously dredged and the sand which accumulates has to be transported away. There have been several dredgers dedicated for this purpose at different times. During the tsunami, the entire harbor and beachfront was devastated. Post-Tsunami it was cleaned, re-dredged and some facilities repaired by USAID assistance, as reported. But, any celebration of this work was premature as was clear when research was in progress, during 2007-2008. At that time, the harbor was usable for the launch of multi-day trawlers but did not have the displacement depth needed to re-dock the boats after they returned laden with their catch, due to continuous re-silting. Multi-day trawlers leaving Fishtown therefore had to return to Bluebay, another habour down the coast, to unload their catch. The price of fish in Fishtown had therefore, quite ironically, risen, even though the number of multi-day trawlers in Fishtown had multiplied dramatically post-tsunami. In fact, not only had the number of trawlers risen but also their concentration, with entrepreneurs who had been single boat owners before now owning several boats. Since all of the multi-day trawlers are owned by local upper middle class entrepreneurs who re-capitalized several fold after the tsunami, they would have little difficulty sending their boats back out from Bluebay if the Fishtown harbor does not function, since they would have the capital, and networks that extend down the coast.

---

[34] Japan International Cooperation Agency, (Activities/Sri Lanka) accessed on 2 June 2007 at http://www.jica.go.jp/srilanka/english/ activities/data_GA.html.

Additionally, it is important to understand, the dredging of the harbor itself cuts across the political economy of fishing since the sale of sand provides a lucrative income. Procedurally, the income of the sale of sand should accrue to the harbor and be used for the cost of dredging, but there are kickbacks from sand sales. If the harbor did not need to be dredged at all, sand sales would cease, and those engaged in the transportation and sale of sand would lose out. No dredging thus is also disadvantageous, and so, it is predictable that the news report will lead to an inquiry and dredging may well re-commence.

Complicating the matter further is the lucrative but illegal trapping, cleaning and exporting of jelly fish – certain of its body parts having a high value as a medicinal product in some East Asian markets – which are processed right on the beach. While this work gives off a terrible stench, it also provides a steady income to many in the town. The accessibility of the harbor to multi-day trawlers does not seem to have a bearing on jelly fishing; in fact, a very active harbor may well be a hindrance.

Given this complex position of the harbor as a node in the cycle of production of several commodities, in different cycles of value, its place in the political economy of Fishtown cannot be represented by a single 'Harbor' Committee. Such committees, which pass for the 'community' in the discourse and practice of aid, are, as I have been arguing, hierarchically and politically constituted. In Fishtown, the local layer of capitalists who have an entrepreneurial interest in fishing are linked, often by long associations with powerful national politicians, who in turn are allied with officials of state agencies that oversee harbor management and fisheries nationally. These capitalists are both Sinhala and Malay, allied by links to electoral political parties and national leadership, as previously noted. The Malays, though a majority in this town, operate under a hegemony of Sinhala culture; for example, parts of the community celebrated the Id with public festivities that included all the 'play' and 'games' of the Sinhala New Year. They are adept, as are all the others in the settlement, in forming committees, associations and societies, in having meetings, working with plans and grants and producing a self-presentation of a kind of unified 'community' that satisfies INGO donors and national-level agencies that work with donors, in this case, the CFHC.

The Seaville/Greenwood Bridge

An even more complex social argument obtains in the Seaville/Greenwood zone, in relation to aid and 'community.' As cited earlier, it has been claimed that the US$ 10.3 million bridge now completed as part of an USAID project was, "developed in consultation with the communities; women and all ethnic and religious groups were represented."[35] Again, an examination of the history of planning proposals here would be edifying. The bridge was part of an earlier plan of a much larger infrastructural complex, which included an inward curving highway leaving large beach-front spaces open for highly-capitalized tourism. This resource development plan which was proposed after the tsunami by several agencies of the state, led to widespread opposition and street protests with existing hoteliers in the vanguard, in June 2005. This was amplified by the Movement for Land and Agricultural Reform (MONLAR), a Sri Lankan NGO.[36] The protests were portrayed in the international press[37] and by the celebrity political analyst Naomi Klein who visited Sri Lanka under the auspices of MONLAR, as one between the 'people' and rapacious 'disaster capital,'[38] which returns us to the naïve understanding of 'community' I have discussed above. It is worth noting that the 50-60 small hotels and guesthouses that were destroyed in the tsunami, were during the time of the protests, being already rebuilt with infusions of petty overseas

---

[35] USAID, *op. cit.*

[36] *Justice to the Tsunami Victims* (Submission to the European Parliamentary Committee on Development), MONLAR pamphlet (n.d.): 7-11.

[37] "Sri Lankans protest as officials seize their land for tsunami 'safety zone'," *The Independent* (London), 10 June 2005.

[38] Naomi Klein, *The Shock Doctrine: The Rise of Disaster Capitalism* (London: Metropolitan Books, 2007), pp. 389 & 385-405, *passim*.

capital. By mid 2006, when research for this project began, well over half the hotels, guest houses, cabanas and restaurants that dotted the seaside stretch of the old road had European or Australian partners who had provided capital for reconstruction and supervised or partnered the management of these properties with Sri Lankans. Very few establishments were wholly owned by Sri Lankans, and of those, none by families that had lived in the settlement for several generations. The overseas capital concerned could be traced, in most instances, to visiting surfers who had formed relationships with Sri Lankans over years of visits and sometimes formed affinal alliances. In a curious irony, the conflict over the proposed post-tsunami redevelopment was between large scale Sri Lankan capital, that had major holdings in other tourist destinations in the country and outside such as South India and the Maldives and on-going alliances with European tour operators, on the one hand, and small Euro-Australian capitalists allied with petty bourgeois Sri Lankan managers. The majority of these managers were Muslim, a few Sinhala;[39] those who were selected as leaders had much experience in numerous, previously-formed associations and societies concerned with rural development and mosque management.[40] "Community" mobilization was again successful and plans for the large-scale tourist development have not been implemented. So while the infrastructural elements that would have accompanied the bridge have not materialized, the bridge itself has not been scaled down, making persuasive Kath Noble's argument that it is a "A Bridge Too Far."[41]

The main local beneficiaries of the USAID infrastructural project are a small kin group in Greenwood, 20 kilometres south of Seaville, who control the Buddhist temple (founded in the 1920s together with the Rural Development Society), several Grama Niladhari

---

[39] The one Tamil entrepreneur who remained, however, recalled far more Tamil inhabitants in the settlement, in the years before STF (Special Task Force) repression, which began in the mid 1980s.

[40] For a similar point about leadership and protest, see Malathi de Alwis, this issue.

[41] Kath Nobel, "A Bridge Too Far," *The Island* , 09 July 2008.

positions and representation in a local government institution, the *Pradeshiya Sabhawa*. This kin group, which is part of a larger group considered 'kavalam' or 'mixed' since at least the 1950s and now called '50-50' because of their composition of inter-married Sinhala and Tamil families of high ranked castes, are able to produce their own interests as those of the 'community.'[42] Having secured contracts to supply unskilled labor for the bridge building, this 'community' leader supplied labor, drawn from the low-ranked Tamil caste groups resident in the settlement, none of whom have ever been able to secure a permanent government job. These laborers were paid half of what they were supposed to get while the rest, it has been suggested, was skimmed by the contractors. This claim was given credence, by the multi-storey mansion replete with ornate pillars which is in the process of being built for the kin group leader, deep in the heart of Greenwood.

But what is more fascinating is that while the bridge over the Seaville lagoon and several other infrastructural projects have been built in Greenwood, including a large fish auction house and wide, paved roads that stretch for miles along the deserted beach, the somewhat explicit arguments of some members of the dominant kin group, like the temple monk, that the settlement is quite content in its isolation, have in fact won out as the authoritative view of the 'community,' since Greenwood remains as isolated as it was before the tsunami. It is one of the few places in the country that is connected to the larger national transportation network by just one road which runs through a flood prone plain. It is unlikely, given the views of the 'community,' that there will be much articulated support for the strengthening of this transportation link.

Conclusion

'Community' 'consultations' have become naturalized in aid discourses and practice produced by INGOs and their local partners

---

[42] For an ethnographic account of the community in the 1950s, see Yalman, *op.cit.*, pp. 310-324.

and are intertwined with other naturalized categories like empowerment and participation. While I am in agreement with critics, such as
Stirrat, who have traced the Christian and Orientalist origins of these
categories, it has been my contention in this paper that these categories
have been appropriated by a nationalist politics that has made 'community' its own and has come to see 'participation' as part of a process
of 'reform,' intertwined with its own Protestant practice. Nevertheless,
these practices of 'community' are of course also practices of the hierarchical reproduction of social order and the making of an authoritative
'voice', which is predicated on doxic silencing and repression. It would
be naïve to call for more or better 'empowered' communities as an an-
tidote to this – the complexity of the social world will not yield to pla-
titudes.

Pradeep Jeganathan is a consultant Social Anthropologist who
lives and works in Sri Lanka. He has written extensively on nationalism, and perpetration and survival of violence and is co-author of the
2007 *Encyclopaedia Britannica* anchor article on Anthropology.

Finding Gampöng: Space, Place & Resilience in Post-Tsunami Aceh[1]

Saiful Mahdi

Abstract

In this paper, *gampöng*, a spatial and cultural concept of community in Aceh, is revisited through a discussion of resilience in post-conflict and post-tsunami Aceh. This paper shows how Acehnese use their social relationships and social networks to cope with calamities. To examine this social and cultural capital, two communities in peri-urban Banda Aceh were observed both during displacement and re-settlement in their original villages. Social cohesiveness prior to disaster, leadership during emergency period and afterwards, as well as interaction with outside intervention in the context of emergency as well as rehabilitation programs and projects were among the main factors that shaped the formation of a 'new' community in a 'new' settlement. Whether these new communities and new spaces become places that can be called *gampöng* will depend on how these communities utilize their new environments and infrastructure for their social interactions.

[1] I would like to thank Eva-Lotta Hedman and Malathi de Alwis for their leadership in the research project, on which this paper is based, as well as their comments on initial drafts of this paper. I am also grateful for comments from fellow researchers Vivian Choi, Jacqueline Siapno and other participants at the December 2007 conference in Banda Aceh at which I shared the preliminary results of this paper. The remaining errors are mine.

Introduction

In the wake of the 26 December 2004 tsunami, Aceh has seen various interventions across its affected urban and rural communities. Since the early intervention during emergency period, however, it was not clear as to what the appropriate unit for intervention should have been: district, sub-district, village, family or individual survivor. While governmental and non-governmental coordination has existed at various levels, humanitarian interventions seem to have been concentrated at the sub-district (*kecamatan*), village (*desa, kelurahan, gampöng*), and family level. It is important and interesting, therefore, to see how these interventions have played out at those levels, and at the village level, in particular.

One assumption for such observation is that Acehnese are very resilient when they are united as a community, at a *gampöng* level. One, thus, might wonder what the roots, if any, of this resilience are. If the people can survive the conflict for years without outside intervention other than that of limited support from its diaspora, there must be some kind of "social and cultural capital" within the communities. If so, *gampöng* as a social network and structure, no matter how it was weakened during the conflict, I argue, plays a pivotal role in Acehnese's resilience to survive both the conflict and the tsunami.

That said, it is important to note that Aceh never isolated itself from the outside world. Quite to the contrary, Aceh and the Acehnese have a long historical record as an open and cosmopolitan society. It was a major hub along the international trade routes of Malacca Straits connecting the Indian Ocean and Southeast Asia.[2] It was the last harbour for Hajj pilgrims to Mecca for Muslims in the archipelago, which gave it the name "The Veranda of Mecca". The Dutch war, the Darul Islam rebellion and the contemporary conflicts with its martial law

---

[2] As nicely put in Anthony Reid, "The Cosmopolitanism and Uniqueness of Aceh," speech at the Aceh Cultural Institute, Banda Aceh, 4 June 2005. For a more formal account, see A Reid, *An Indonesian Frontier: Acehnese and Other Histories of Sumatra* (Singapore: Singapore University Press, 2005).

which closed Aceh from the outside world, therefore, can be seen as exceptions interrupting such long-standing linkages and networks between Aceh and the outside world.

In this study, I seek to answer the following questions: (a) What are the impacts of outside interventions to Acehnese social structure, especially in relation to the concept of *"gampöng"* (socially and spatially) in post-tsunami Aceh? (b) How are outside interventions helping/not helping post-tsunami Acehnese communities compared to the way Acehnese helping/not helping themselves through their social relation. In particular, I seek to understand how communities in post-conflict and post-tsunami Aceh regard the new "space" of living after new houses, roads, and other infrastructure have been rebuilt. Can the new "space" become their new "place"? That is, a "space" with "soul", "non-secular" and which enables the previous or new social structure, based on Acehnese "cultural *gampöng*", to flourish.

Methodology

To understand how new spaces have emerged in post-tsunami Aceh and how it relates to the notion of *gampöng,* this paper turns to a comparative investigation of two communities at the village level and the relocation challenges they faced in the post-tsunami period. The first community observed, the Al-Mukarramah community in Punge Jurong village in Banda Aceh, was effectively divided in the early post-tsunami relocation as the entire village was not accommodated into a single barrack camp, nor subsequently entirely reunited. In contrast, the second village under study here, Lambung village, also in Banda Aceh, remained united, having rejected (temporary) resettlement into different barrack camps and subsequently succeeded in moving back into their village. This comparison supports the argument advanced in this paper that united communities were better able to negotiate outside interventions, including with international organizations while others split vis-à-vis solutions offered, thus could not keep *gampöng* together.

Both qualitative and quantitative approaches were used to try to understand the role of *gampöng* in Acehnese's interaction with outside interventions, especially in the case of the two villages. Interviews with *gampöng* key stakeholders like the village head, *imam*, neighbourhood chiefs, women's groups, community organizers, youth groups were done in the two villages in Banda Aceh: Lambung and Punge Jurong. Questionnaires were used to complement the lack of village statistics to help substantiate some quantitative measures like demography before and after the tsunami, the livelihood situation, etc. This way, data collection will also be based on the collective memory of the inhabitants of the two villages.

Gampöng, the Acehnese village

The smallest unit of communities (and territory) in Acehnese society was a *gampöng* until Indonesia under Suharto tried to fit all community structures into *desa* or *kelurahan* with Law No. 5/1979. *Desa* and *kelurahan*[3] are Javanese terms for village structure, popularized across Indonesia by Suharto's administration. Both are territorial administration and population units under a sub-district (*kecamatan*), a difference being that a *desa* can administer itself and its leader (*kepala desa*) is elected by the villagers while a *kelurahan* cannot administer itself and its leader (*lurah*) is a government-appointed employee, not elected by the villagers. Since then, more villages in urban areas have become *kelurahan* while some maintain its *gampöng* status with less "sovereignty". In rural areas, however, most remain a *gampöng* although with decreased capacity under the state administration system. But, as a *gampöng* was not recognized in the administration, it was considered the same as *desa*.

While the status or designation of a village is not trivial, at least in securing state supports, a broader issue in this study is village as 'sociality', ways of being, and sovereignty as depicted by a *gampöng* in the

---

[3] For the definition, see *Statistics Indonesia* website accessed on 28 October 2008 at http://www.datastatistik-indonesia.com/content/view/928/950/ .

Acehnese psyche. For example, a *kelurahan* is led by a *lurah* appointed and paid by the government, while a *gampöng* is led by a *keuchik* elected by the *gampöng*'s inhabitants. A *keuchik* gets some support from the government but a *keuchik* is not a government employee. Although a *desa* also indicates certain "sovereignty" in that its leader is also democratically elected, it is different from a *gampöng* in the Acehnese cultural context. A *gampöng*, for instance, is also nested under a *mukim* with several other *gampöngs*. That is why, a *gampöng* lost its relative strength as a *mukim*, a higher unit, groups of *gampöngs*, was abolished by the government and replaced by *kecamatan* (sub-district) led by a *camat*, a government trained and paid employee. It was the case until recently when the approved Law No. 11/2006 on Aceh Governance (Undang-Undang Pemerintahan Aceh [UUPA]) took effect. The new law, a manifestation of the Helsinki Peace Memorandum of Understanding (MoU) of 15 August 2005 between GAM (The Free Aceh Movement) and the Indonesian Government, would let the Acehnese revert to its former governance structure based on *gampöng* and *mukim* within two years after the law was decreed. Today, many *kelurahan* or *desa* in Aceh have gradually transformed themselves back into *gampöng*.

This shift in legislation raises the question of what constitutes a *gampöng* in terms of sociality, rather than simply legislation, in Aceh today. How significant is the *gampöng* in the lives of people in Aceh? Has it changed in the aftermath of conflict and tsunami? As suggested by this paper, new post-tsunami settlements built by multi-donor funded reconstruction projects may provide a "space" with rows of neat or not-so neat houses. Beyond such reconstructed spaces, however, a *gampöng*, this paper shows, is also a "place" with a "soul" connecting its inhabitants to one another *and* with the space. As Steve Harrison and Paul Dourish put it

> Whereas space refers to the structural, geometrical qualities of a physical environment, place is the notion that includes the

dimensions of lived experience, interaction and use of a space by its inhabitants.[4]

As noted by Snouck Hurgronje, a *gampöng* (Malay, *kampung*), in spatial terms, is "the smallest territorial unit" in Aceh.[5] In his writings, Hurgronje also offered rich descriptions of a *gampöng* as a place steeped in distinctive forms of lived experience, interaction and use of space:

> There are the courtyards, part of which are utilized as gardens, containing one or more houses separated from one another and from *gampöng*-path (*jurong*) by fences; then the whole *gampöng* surrounded by a fence of its own, and connected by a gate with the main road (*rèt* or *ròt*) which leads through fields and gardens (*blang* and *lampoih*) and tertiary jungle (*tamah*) to other similar *gampöngs*.[6]

Hurgronje noted that *gampöng* first emerged as a unit formed around kinship groups (*kawoms*) or sub-divisions of such a group, "which added to its numbers only by marriages within its closure, or at most with the women of neighbouring fellow-tribesmen."[7] Traditionally, each *gampöng* has had a *meunasah*[8] "as a nightly resting-place of all the full-grown youths of the *gampöng*, and all the men who are tempo-

---

[4] S Harrison and P Dourish, *Re-Place-ing Space: The Roles of Place and Space in Collaborative Systems* (Xerox Palo Alto Research Center, Rank Xerox Research Centre, Cambridge Lab (EuroPARC), 1996): 67-76.

[5] Snouck Hurgronje, *The Achehnese*, trans. A W O S Sullivan (Leyden: late E J Brill and London: Luzac & Co, 1906), p. 58.

[6] *Ibid.*, pp. 58-59.

[7] *Ibid.*, p. 59.

[8] "This word which also appears in the form *beunasah, meulasah*, and *beulasah* is derived from the Arabic word *madrasah*, meaning a teaching institute...." (Footnote in Hurgronje, *op. cit.*, p. 61)

rarily residing there and have no wife in the *gampöng*."[9] A *meunasah* also provided a place for Muslim prayer, both individually and in congregation as well as for meetings to discuss matters of public interest. Thus, a *meunasah* was inseparable from a *gampöng*. *Gampöng x* is sometimes also called *meunasah x* instead. A *meunasah*, therefore, can mean a physical structure for prayer or a territory (*gampöng*). That is why some have referred to Acehnese society with its *gampöngs* as a 'religious territory society'.[10]

Traditionally, a *gampöng* is headed by a *keuchik*[11] (village head), who used to be chosen from among the *uleebalangs* (chief) of the *kawom* or their successors. Later, however, as *kawoms* grew larger and the boundaries among them became increasingly blurred, the *keuchiks* came to be chosen more democratically. Since 2007, one year after the Law of Governance of Aceh (LoGA), Acehnese in *gampongs* have been electing their *keuchiks* directly through *Pilchiksung* (*pemilihan keuchik secara langsung* or direct *keuchik* election).[12] A *meunasah* has also had a role as a teaching institute of sorts, mostly focused on religious affairs, and, as such, has been under the leadership of a *teungku* or *imeum meunasah*. Thus, a *gampong* should be administered jointly by *keuchik* and *teungku*, each according to their own specified duties. In one *gampöng*,

---

[9] Hurgronje, *op. cit.*, p. 61.

[10] Nya' Pha, *Imeum mukim, keberadaan dan peranannya kini: suatu penelitian di Kecamatan Seulimeum, Aceh Besar: laporan penelitian* (Role of Imeum Mukim, a village traditional leader in Kecamatan Seulimeum, Kabupaten Aceh Besar, Aceh Province), research report (Darussalam, Banda Aceh : Universitas Syiah Kuala, 1998) quoted in Sulaiman Tripa, "Memahami Budaya dalam Konteks Aceh (Understanding Culture in The Aceh Context)," (16 May 2006) accessed on 10 December 2006 at www.acehinstitute.org.

[11] Some use *geusyik* or *geuchik*. Local Law (Perda) No. 7/2000 uses *geusyik*, Qanun No.5/2003 uses *keuchik*. Thus *keuchik* is used in this paper.

[12] For a brief account on this, see Sujoti, A. and F. H. Rahman, eds., *Membangun Aceh dari Gampong: Catatan Ringan dari Riset Monitoring Pemilihan Keuchik Langsung (Pilciksung)* [Building Aceh from the Gampong: Notes from Research and Monitoring on Direct *Keucik* Elections (*Pilciksung*)] (Yogyakarta: Institute for Research and Empowerment (IRE), 2007).

there might be more than one *meunasah*. According to Qanun No. 5/2003, the *keuchik* is the executive of the *gampöng* in day to day *gampöng* governance. The *Keuchik* works together with *tuha peut* (the elder four), *teungku* or *imeum meunasah* and village secretary.

A *keuchik*, from the perspective of a *gampöng*, is not only a leader for its people and territory, but also the caretaker of *adat*, the customary law. *Tuha peut*, the elder four, are the four persons considered knowledgeable and resourceful within the population, who help the *keuchik* as wise-men, advisory council, and in some cases as judge.[13] "They are men of experience, worldly wisdom, good manners, and knowledge of *adat* in the *gampöng*.[14] Syafei Ibrahim, a Acehnese scholar, wrote that "authority in Aceh derives from a variety of sources, including supernatural and spiritual powers (*kesaktian*), heredity (*keturunan*), knowledge (*ilmu*), and a combination of personal characteristics including wise and just (*adil dan jujur*), courageous and decisive (*berani dan tegas*), generous (*dermawan*), kind and hospitable (*ramah tamah*)."[15] Craig Thorburn noted that "while popular imaginations envision a *Keucik* as protecting and upholding the interests of his community, historically, *Keucik* have acted as the agents of higher authorities (originally *Uleebalang*, and more recently, district and national government)."[16]

Today, a *keuchik* is elected democratically by people in the *gampöng* while *lurahs* are appointed top down by the government through

---

[13] Nya' Pha, *op. cit.*, in Tripa, *op. cit.*

[14] Hurgronje, *op. cit.*, p. 75.

[15] Syafei Ibrahim, "Kewibawaan dalam Pandangan Masyarakat Aceh" (Authority in the Viewpoint of the Acehnese), Journal *Ilmiah Administrasi Publik* VI, no. 1, (2006) accessed on 2 May 2008 at http://publik.brawijaya.ac.id/?hlm=jedlist&ed=1125507600&edid=1135589533 in C Thorburn, "Village Government in Aceh, Three Years after the Tsunami," Center for Southeast Asia Studies, Paper CSEASWP1-08 accessed on 2 May 2008 at http://repositories.cdlib.org/cseas/CSEASWP1-08.

[16] *Ibid.*

the *camat* (head of *kecamatan*, sub-district). Therefore, there is a real sense of democracy and sovereignty in a *gampöng*. Additionally, all matters of public interest in a *gampöng* are discussed openly and decisions are made based on *mupakat* (*muwafakat*, Arabic). Decision by palaver has to be made at least by the three officers in a *gampöng*: *keuchik, teungku meunasah,* and *ureung tuha* or *tuha peut*. Administration of a *gampöng*, thus, is "composed of these three elements."[17]

In a *gampöng*, there are also "professional" customary offices to help and in consultation with the *keuchik*. They are village elders considered the most knowledgeable and wise in each field of public affairs. The *panglima laöt*[18] helps the *keuchik* in matters related to sea, if a *gampöng* is next to one. The *peutua seuneubok* deals with forestry, gardens, andun-irrigated agricultural field; the *keujreun blang* helps the *keuchik* regulate water and farming in the rice field; the *haria peukan* helps maintain order, security and cleanliness in the village market. The *haria peukan* also collect dues from sellers and vendors. The *panglima laot* is usually assisted by the *syahbanda* who lead and regulate boats and traffic in a river or harbour.

When the office of the *keuchik* was incorporated into the Indonesian government structure in the 1980s, John McCarthy found that the role of village elders was diminished while the *keuchik*'s executive power increased. There was also an increasing separation of powers between state authority and customary and/or religious authority.[19] Thorburn, whose studies tried to look at post-tsunami reconstruction

---

[17] Hurgronje, *op. cit.*, p. 64

[18] For the role of the *panglima laöt* and Aceh sea customary law (*hukôm adat laöt*) after the tsunami, see, M Adli bdullah, S. Tripa and T. Muttaqin, *Selama Kearifan adalah Kekayaan: Eksistensi Panglima Laôt dan Hukôm Adat Laôt Aceh* (Banda Aceh and Jakarta: Lembaga Hukôm Adat Laôt Aceh/ Panglima Laôt Aceh and Yayasan Kehati, 2006).

[19] J McCarthy, "Village and State Regimes on Sumatra's Forest Frontier: A Case from the Leuser Ecosystem, South Aceh," RMAP Working Papers, no. 26 (Canberra: Resource Management in Asia Pacific Project, Research School of Pacific and Asian Studies, Australian National University, 2000).

from a *gampöng* perspective, wrote, "many popular, media and donor descriptions give the impression of the *Keucik* as the wise and trusted keystone of Acehnese village society."[20]

The Case of Punge Jurong and Lambung

Gampöng Lambung and *Kelurahan* Punge Jurong are two among sixteen villages in thesub-district of Meuraxa, City of Banda Aceh, in Aceh, Indonesia (See Figure 1). During the conflict, like many other communities in Banda Aceh, the two villages have been relatively free from the conflict tension. While both have presumably been the host for conflict Internally Displaced Persons (IDPs), there is no exact account of number and information of these IDPs.[21] It might be assumed, however, that Punge Jurong had hosted relatively more conflict IDPs than Lambung due to the fact
that Punge Jurong had a larger migrant population from conflict prone regions like Pidie and North Aceh. The two villages, like all those in the Meuraxa sub-district, were totally destroyed by the tsunami of 26 December 2004.

Gampöng Lambung, situated about 1-2 kilometres to the east from the sea line of Banda Aceh, was almost "cleared" by the tsunami wave leaving no physical structures intact. Across its 31 hectares area, there was only one house spared, but badly battered by the tsunami.[22] The villagers are mostly native and deeply rooted to their spatial territory. Many of the inhabitants in Lambung were (and still are) related to one another by family ties (*kawôm*) or marriage. These two characteristics seem to have made social ties stronger in this village, resulting in, for example, easier and faster consensus making after the

---

[20] Thorburn, *op. cit.*

[21] Interview with Mr. Abubakar Ishak, Al Mukarramah neighborhood head of Punge Jurong on 21 February 2007, and Mr. Hardiansyah, one of the community organizers in Lambung, 22 February 2007.

[22] The villagers of Lambung, with the permission from the owner, decided to keep the remains of the house as it was as a tsunami monument (Interview with Mr. Masykur, 38 years, of Lambung in Lambung 3 December 2007).

tsunami. The village, for instance, reached an agreement for land consolidation in a relatively easy consensus. The village is headed by Mr. Zaidi M. Adam, a native of Lambung with a five-year college degree, who lives there and was directly elected by the villagers.

Prior to the tsunami, people in Lambung had various occupations. Many worked in home industries producing traditional sweets for which Lambung was famous. Some others work as fishermen, fish breeders, government employees, businessmen, small traders, carpenters, or private employees. Most of the villagers in Lambung have at least a high school education or higher.[23]

Kelurahan Punge Jurong, on the other hand, is more diverse in term of nativity of its population. A significant number of its population, more than 60 per cent[24], is from Pidie and North Aceh, a conflict hot spot. The tsunami wave brought hundreds of corpses and debris from the other closer-to-coastal villages including those from Lambung, two-village away from Punge Jurong. Punge Jurong itself is about 3-4 kilometres to the east from the sea line, covering the area of 42.2 hectares. The remaining structure of houses and other buildings, about 15 per cent of total buildings in this village, made it difficult to do, for example, land consolidation and infrastructure reconstruction. Punge Jurong is led by Mr. Faisal (no last name), who is not a native nor living in Punge Jurong. Faisal was appointed by the *Camat* (subdistrict head) after he finished a four year college at STPDN (*Sekolah Tinggi Pemerintahan Dalam Negeri*), a special school run by the Interior Minister in Jakarta to train state employees designated to lead *desa, kelurahan,* and *kecamatan.*

---

[23] Government of Lambung, "Rekonstruksi Gampong Lambung," presentation made in several occasions, including in front of potential donors and village meetings.

[24] Interview with Faisal, STTP, *Lurah* (head) of Keluarahan Punge Jurong in Banda Aceh, 14 December 2006.

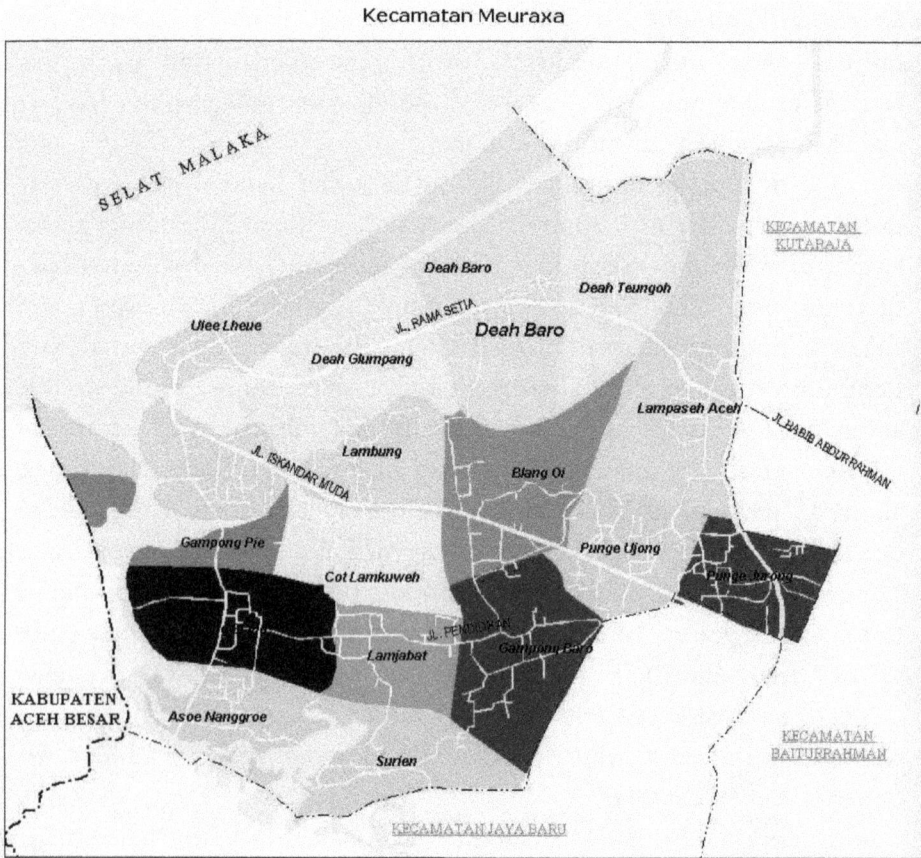

Figure 1: Map of Meuraxa Sub-District (Kecamatan), Banda Aceh, Aceh

A demographic comparison of Punge Jurong and Gampöng Lambung is shown in Table 1. Punge Jurong as a whole is much larger than Gampöng Lambung in term of population. Punge Jurong had a population density of 2,510 people per square kilometre before the tsunami, while Lambung had 385 people per square kilometre.

Village	Before Tsunami				After Tsunami									
					Missing			Dead			Survivor			
	H	M	F	T	M	F	T	M	F	T	H	M	F	T
Punge-Jrong	1122	2999	2950	5949	38	215	253	1691	2070	3761	778	1570	865	2435
Lambung	276	683	558	1241	18	55	73	462	396	858	149	210	110	320

H=Household, M=Male, F=Female, T=Total

Table 1: Population Comparison between Punge Jurong and Lambung[25]

The size of Gampöng Lambung's population is very similar to Al-Mukarramah neighbourhood, the largest among five neighbourhoods in Punge Jurong. This is one of the reasons that this neighbourhood is used in this case study, instead of the whole of Punge Jurong.

Given its proximity to the central market, more inhabitants in Punge Jurong—especially those living in Al-Mukarramah neighbourhood are merchants and small traders. Only a small fraction of its population worked in other fields such as being state employees, teachers, private employees, and home industries. This might have made the people in Al-Mukarramah more independent minded.

Having followed the two communities of these villages, from their temporary shelters as IDPs at the aftermath of the tsunami to their original villages, I observed that the social structure in the two vil-

---

[25] Dr. Tarmizi Yahya, MM, head of Meuraxa Sub-District (Kecamatan), presentation at Komite Rehabilitasi and Rekonstruksi Meuraxa (Korrexa) meeting (2006).

lages have contributed to the different levels of success in reorganizing and reformation, thus, resilience, of these communities.

Navigating Displacement: A Closer Look

The cases of Al-Mukarramah neighbourhood of Punge Jurong village (*Kelurahan* Punge Jurong) and Lambung village (*Gampöng* Lambung) are reported here to show how outside interventions could influence the dynamics within a community and thus determine the reorganisation of a *gampöng* after the tsunami. Disorganised provisions of relief supplies, livelihood options, and power struggle were among major causes of social strife in the post tsunami communities or, at the very least, ones which might trigger the problems.

Al-Mukarramah: The Problem of Split-Community

When the Indonesian government decided to provide barracks, military style temporary shelters, or temporary living centres (TLCs), for tsunami IDPs, the challenge to Acehnese social structure and cultural practices was very significant. [26] Not only are the barracks culturally inappropriate, their provision is also not enough for every IDP causing split-family and split-community problems. Because the units of barracks built were not enough for every IDP or were slowly built for people to move in, villagers were divided in several groups to fit in to several barracks locations or just to fulfil a barrack quota, which was up to five or six persons in a room of four by six meters. Each unit of barracks consisted of twelve rooms, so that authorities aimed at moving about sixty persons into a unit of barrack. There were reports that families might be divided into units in barracks or combined with

---

[26] Two year after the tsunami, Oxfam reported that about 70,000 are still living in military-style temporary shelters called "barracks" in Aceh, while more are still living with host communities throughout the region. See, "Oxfam calls to step up response for 70,000 tsunami survivors living in barracks in Aceh" accessed on 10 January 2007 at http://www.oxfam.org/en/news/pressreleases2006/ pr061116_aceh.

other IDPs. The barracks are not so friendly to vulnerable groups like women, children, the elderly, and the disabled.[27]

From a more macro point of view, the division of IDPs from the same village into different barracks location is more troubling. There are cases where some portions of IDPs from the same village were moved into barracks while the rest were not due to limited number of barracks the government could provide. In both cases, the cohesiveness of local social structures (*gampöng*) is under pressure. Moreover, it is difficult for the concept of *gampöng* to exist and be nurtured in those shelters. People find their "space" in the barracks, but seemed to miss their "place". That is why many used these TLCs as a place to mark their existence only for the provision of relief supplies. Many were absent from life in the barracks and chose to live somewhere else, but they would come to the barracks to get supplies on designated days.[28]

Al-Mukkaramah neighbourhood is one of five neighbourhoods in the village of Punge Jurong, Meuraxa sub-district, City of Banda Aceh. Although situated nearly four kilometres inland, the neighbourhood could not stand the force of the tsunami wave. Prior to the tsunami, the neighbourhood had 3812 members (382 households), but at the end of 2007, the population is numbering around 942 only.[29] Until June 2006, some 192 of the survivors were living in Lhong Raya barracks; about 165 were living in barrack units and tents within the neighbourhood, while the rest were scattered in different host commu-

---

[27] Reports show that barracks were not in compliance with SPHERE. . Reports from women advocacy groups are especially critical to daunting living condition in the barracks. See also, E E Hedman, "Back to the Barracks: Relokasi Pengungsi in Post-Tsunami Aceh," *Indonesia* 80 (October 2005):1-19 and L Age, "IDPs Confined to Barracks in Aceh," *Forced Migration Review*, Special Issues nos. 22-23 (2005).

[28] Interview with Ismail Husen, resident of Punge Jurong, 15 December 2006.

[29] Author interview with Abubakar Ishak, Head of Al-Mukarramah neighborhood, Punge Jurong, 14 December 2007.

nities/families in Banda Aceh, Pidie, and North Aceh. Some went to as far as Medan, Jakarta and Malaysia.[30]

Most of the residents in this neighbourhood came from Pidie. Famous as traders, they based their livelihood in the central market of Banda Aceh as road side vendors, shop owners or bigger traders (distributors). In the days after the tsunami, survivors from this neighbourhood were scattered in several IDP camps in Banda Aceh. Some went back to Pidie to drop their smaller children, women or the elderly to their original *gampöng*.

When they got back to Banda Aceh in early to middle of January 2005, to look further for missing family members, or to find humanitarian and relief supplies, they were based in a public building named *Gedung Sosial* near the gate to the city of Banda Aceh. Led by Abubakar Ishak (49 years), a community organizer, asked by the group to be in charge of replacing the missing neighbourhood head, they seemed to be very united during the first months after the tsunami.[31] Problems and disputes usually occurred with other groups of IDPs from other villages in the same shelter where they lived and competed over relief supplies and aid distributions.

However, when TLCs in the form of barracks were built by the government in April-May 2005, and the number of units for IDPs from this neighbourhood was not enough, the division started to occur. The trigger was the problem of who got to move to the barracks and who would stay in the public buildings. Fortunately, around that time, the government decided to let the IDPs to return to their villages cancelling the issue of buffer zones in Banda Aceh. Mr. Ishak led meetings with his people and decided to prioritize women and women with children plus some male IDPs to move to barracks in Lhong Raya of Jaya Baru sub-district, about four kilometres from the neighbourhood and the public building. The rest voluntarily decided to return to the

---

[30] Author interview with Abubakar Ishak, Head of Al-Mukarramah neighborhood, Punge Jurong, February 2006.

[31] The author visited the shelter several times in January-April 2005.

destroyed but already accessible neighbourhood in Punge Jurong. After that, what the author usually heard were jokes among the two groups; the groups who stayed in the barracks and the groups who returned to the neighbourhood and lived in tents and the remains of houses, teasing one another about their living conditions.

The groups who returned said that they were lucky in deciding not to insist on moving to barracks as the condition in barracks were terrible with limited yet bad water and sanitation facilities, lack of privacy leading to feeling of dehumanization, and so forth. Some coming from barracks admitted the situation but claimed that they can be more restful in a close unit of TLC rather than living in the open before moving to barracks. They admitted the initial poor sanitation and water problem in barracks. On the other hand, people from barracks teased the early returnees as having to cope with worse conditions on the ground.

During the emergency period following the tsunami, humanitarian and relief organizations, especially those of international ones, held back their interventions to the early returnees and barracks inhabitants. For the first, they were not sure whether the Indonesian government would allow returnees to coastal areas while for the latter they knew that barracks are not *Sphere* compliance.[32] With no or limited interventions, the IDPs confined to barracks[33] and those of early returnees had to live under grim living conditions on their own ground. But, the jokes and the teasing of one another kept going on as part of their everyday life.

The situation became more serious after the government greenlighted the right to return for early returnees and asked international organizations to step in to help the IDPs in barracks despite the breach

---

[32] *Sphere* is the generally accepted Guiding Principles on Internal Displacement, including by the UN

[33] See Age, *op. cit.* and Hedman, "Back to the Barracks," *op. cit.* and E E Hedman, "The Right to return: IDPs in Aceh," *Forced Migration Review*, no. 25 (2006):70 for an account of barracks for IDPs in Aceh.

of international standards. Afterwards, many organizations did step in to help both groups of IDPs. But then the problem of division started to worsen as the geographical divide prevented effective communication of the two IDP groups who were actually from the same village. Early returnees claimed that their fellow villagers living in the barracks were well taken care of as everything was provided by the government and relief organizations. In contrast, the people in barracks claimed that they did not get enough and insisted on getting relief supplies delivered directly to the neighbourhood as well. Suspicions and distrust among one another started to take place and increased as some village figures competed for influence over power and aid supplies.

The issue of the "unit for intervention" was more significant than ever. For example, which unit should be used for relief supplies coordination? The village was already split, the families were scattered outside the village, and the *kecamatan* (sub-district level) was too remote, mentally and physically, to a neighbourhood. This is not to mention cases of people with "vague *gampöng* membership" like Mr. Ibrahim Din, a 55 year-old man, as chronicled by Sorayya Khan, a writer from Ithaca, NY[34]:

> Ibrahim Din is originally from Simpang Mulieng, a hillside village in North Aceh [some 200 kilometers away from Banda Aceh] that consists of 1000 people, about 200 households. He offers a detailed description of life during the conflict and describes his village as being a battlefield between GAM and the military. He is a fish breeder and farmer, but feared for his life sometimes when he visited his farms or fish pools. He had a wife and seven children. One of his daughters was married to a Punge Jurong resident [in Al-Mukarramah neighbourhood] and she, her husband, and child lived in the neighbourhood.

---

[34] The writer is Sorayya Khan, a novelist living in Ithaca, New York. She visited Punge Jurong to do research on the nexus of disaster, memory and trauma. She is also a volunteer and board member of Aceh Relief Fund (ARF). Her trip journal can be found at www.acehrelief.org (accessed 12 April 2008).

He sent three of his other children to stay with her during the conflict because the security situation in his village was so bad and, also, because of the limited educational opportunities available. He lost all four children and his grandchild in the Tsunami.

By approaching the writer who visited the neighbourhood in May 2007, Mr. Din was looking for support to his appeal that he should also be considered for tsunami aid.[35] He argued that although he is not a "direct victim" of the tsunami, nor was he a villager of Punge Jurong, he had all the rights to get tsunami aid like all those who used to live in Punge Jurong and other affected areas. He had lost six family members from the tsunami in the village, including his wife who was visiting the village when the wave arrived, four children and a sixty-day-old grandson. He said that he could only access limited aid provision as he did not live in and was not really part of the village. He had to be insistent in alerting the villagers in Punge Jurong and its authority that a significant part of his family was lost in the tsunami while being part of the village, a *gampöng* in Acehnese. The man was struggling to find membership of the *gampöng*.

The dilemma in Mr. Din's story was actually not unique. There are many other similar stories where people became tsunami victims in Aceh, or otherwise its survivors, due to the fact that they had to flee the interior of Aceh and lived in Banda Aceh and other coastal areas. They became eventual victims in their search for refuge from the oppression of the conflict. They left their *gampöng* in the interior to live in and try to be a member of another *gampöng* in Banda Aceh. How much support they received from their *gampöng* during the conflict and at the aftermath of the tsunami, seemed to depend on how strong their attachment was to a *gampöng* and how well knitted a *gampöng*'s social fabric was.

---

[35] Conversation with Sorayya Khan, Ithaca, NY, 23 May 2007.

Variable	Pre-Tsunami	Post-Tsunami
1. Demography		
Population	3812	942**
Household number	382	276**
Family Size	9,9	3,4
Male	49%*	60,9%
Female	51%*	39,1%
Under 5 years old	n/a	6,1%
Children and adolescent	n/a	16,9%
Elderly (>55yrs)	n/a	5,2%
2. Land and Property		
Own land certificate	67,4%	51%
Own land buying note	14%	2,3%
No land note or certificate	16,3%	41,8%
Average land size	204,54 m2	198,96 m2
Land dispute	2 cases	No cases
Own house	74.4%	51.2%
House renter	7.0%	7.0%
Relatives' house	11.6%	32.6%
3. Livelihood		
Merchants	44.2%	37.2%
Jobless	7.8%	16.3%
Own capital	82.0%	10.0%
Average capital to start	Rp. 28 juta	Rp. 18,5 juta
Income < IDR 500,000	27.9%	34.9%
Income IDR 500,000–1.5 millions	39.5%	34.9%
Income IDR 1.5–3 millions	9.3%	7.0%
Income > IDR 5 million	4.6%	0.0%

Table 2: Summary Statistics of Pre- and Post Tsunami Situation of Al-Mukarramah Neighbourhood, Punge Jurong, Banda Aceh[36]

---

[36] S Mahdi and Cut Famelia Muhyiddin, *Tracking Changes: Where Al-Mukarramah Neighborhood Stands One Year after Tsunami* (Banda Aceh: Yayasan Masyarakat Iqra-UN-HABITAT Aceh and Nias Shelter Support Program (ANSSP), 2006)

The community divide in Al-Mukarramah was also fuelled by the changing demographic and socio-economic character of the neighbourhood. Table 2 shows summary statistics of changes in demography, land and property, and livelihood in Al-Mukarramah neighbourhood. Economic pressure was most daunting because most of the people from this neighbourhood were successful traders with much independence in their lives. Although many had returned to restart their businesses in the central market, they then had much smaller businesses with smaller capital borrowed from cooperative and revolving fund schemes provided by Non-Governmental Organisations (NGOs). Social strife was more latent in this neighbourhood with eroded social capital. Much of this erosion was caused by contentious management and leadership over outside help, to the extent that the community in the barracks outside the village at one point drew their own leadership, causing double leadership when it came to original village affairs.

The culmination of the divide was when *Pak* Abubakar Ishak, the head, decided to resign in early May 2006 in front of a neighbourhood meeting. He said that he was tired of being criticized by his fellow villagers living outside the neighbourhood (barracks and host communities outside the village) who would come back occasionally to the neighbourhood and had not done anything but demanding for supplies and services they already had in their designated temporary shelters.[37] Despite his sincere work to help bring the neighbourhood back to normalcy, some people suspected him of corruption based on hearsay. It was understandable when he said that it was not easy to make everybody happy while they did not even meet one another regularly: some lived in the neighbourhood under tents, barrack, and destroyed houses; some lived in the barracks outside the neighbourhood, some lived far away outside Banda Aceh in their respective original

---

[37] IDPs living in host communities were also provided relief supplies albeit with different satisfactory responses.

gampöng, and yet still some lived in host communities in Banda Aceh and beyond

Variable	Pre-Tsunami	Post-Tsunami
1. Demography		
Population	1.900	400
Household number	320	309
Family Size	6,9	1,29
Male	55 %	60%
Female	45 %	40%
Under 5 years old	n/a	2%
Children and adolescent	n/a	15%
Elderly (>55yrs)	n/a	2%
2. Land and Property		
Own land certificate	40 %	100 %*
Own land buying note	10 %	- %*
No land note or certificate	50 %	0 %*
Average land size		
Land dispute	5 cases	-
Own house	85 %	100 %
House renter	1.0 %	0 %
Relatives' house	14 %	0 %
3. Livelihood		
Merchants	n/a	5 %
Jobless	n/a	30%
Others	n/a	
Own capital	n/a	80%

Table 3: Summary Statistics of Pre- and Post Tsunami Situation of Gampöng Lambung, Kec. Meuraxa, Banda Aceh

This showed that the neighbourhood had become less cohesive after the emergency period. Community solidarity and supportive bonds formed during the emergency period, when people more or less lived together, eroded as they were divided into different "spaces", as communication became more difficult, and public spaces where many could gather together were non-existent. The sense of community was also divided along lines of group loyalty: those from the original village in the interior, from different barracks, those living with host-communities, and those who had returned to their own neighbourhoods. Initiatives and negotiation with outside donors were only on ad-hoc basis taken on by the neighbourhood head. This, for example, had resulted in a very slow house and infrastructure reconstruction in the neighbourhood. The reconstruction in the neighbourhood had not started until mid-2007, two and a half years after the tsunami.

Lambung: A Re-united Community?

More than 1,000 out of 1,300-1,500 residents of Lambung vanished to the tsunami. According to Drs. Zaidi M. Adan, *Keuchik* of Lambung, 100 percent buildings and houses in Lambung were destroyed.[38] Until the end of 2007, there were 320 people registered as official residents in this village, of which about 300 are survivors from the quake and tsunami. The remaining are new residents due to new marriages after the tsunami. *Gampöng* Lambung has been always a *gampöng*, and thus never changed into a *desa* or *kelurahan*. This is the only village with a *gampöng* structure in the sub-district of Meuraxa. Table 3 shows the change of demographic, land and livelihood issues in Gampöng Lambung before and after the tsunami.

Prior to the tsunami, more than 90 percent of people in Lambung were native inhabitants of the *gampöng*.[39] Others came to the village through marriage, not by merely moving into it. Most of the native villagers were related to one another through family ties, fitting a *kawom* concept of the former time of Aceh. This, among others, had always resulted in the strength of the community and social ties among the villagers.

---

[38] Government of Lambung presentation at Aceh Habitat Club, a reconstruction participatory forum hosted by UN-Habitat and the Aceh Institute, Banda Aceh, 29 June 2006.

[39] Interview with Hardiansyah, a native of and community organizer in Gampong Lambung, 15 December 2006.

Fleeing the tsunami, like residents of other villages of Banda Aceh, survivors from Lambung ran and took refuge in the hilly area of southeast Banda Aceh around TVRI (state TV) tower and relay station in *Mata Ie*. After the wave struck Banda Aceh, soon this area became a famous IDP camp with influx from all over Banda Aceh and Aceh Besar. This was also the first tsunami IDP concentration where government and non-government, local, national and international tsunami humanitarian and relief aids gave much helps in the first days after tsunami.

Interestingly, survivors from Lambung did not stay long in the TVRI camp. Instead, they took refuge with their relatives having several houses near the TVRI area. Later, when many more survivors from Lambung got there, they rented a piece of land in the area and erected several tents, allowing them to stick together; separated from the bigger crowd in the TVRI camp but still with fairly good access to aid supplies from the bigger camp. When asked why people from Lambung separated themselves from the bigger crowd of IDPs in TVRI camp, Hardiansyah, one of Lambung's community organizers answered:

> We did not separate ourselves from others. But we centralised ourselves, so we can focus, reorganise, and mobilise our people easily. Everybody who was healthy and strong enough worked together. First, we have these relatives offering us their places to stay, then when it was over-crowded, we rent land close to the place and erected several tents with public kitchen, water pump, and latrine.[40]

They returned to Lambung soon after road access allowed them to do so. They refused to be relocated into barrack units outside Lambung while other survivors from other villages were trying to get one.[41]

---

[40] Interview with Hardiansyah, 29 January 2007, emphasized by author with italics.

[41] IDPs in general were anxious or otherwise indifferent about barracks, eager to welcome the prospect to live in a close unit with provisions at one point, but feeling de-

Instead, they were proposing to donors that were doubtful then, to help early returnees like them provide barracks on site of their village in Lambung. Before they were helped by any donor, they built their own one barrack unit through *gotong royong*.[42] Survivors used salvaged materials and bought new ones with donations from their wealthier fellow villagers in Banda Aceh and elsewhere. Later, an international organization built two more barrack units in the village allowing more survivors to return to Lambung.

There were survivors from Lambung who did not stay together, but all agreed that there was only one "Posko"[43] for every survivor from Lambung to turn to when they needed it; unlike survivors from other villages who opened several Poskos depending on where they took refuge. Survivors from Lambung did spread into different host communities; some even temporarily left Banda Aceh. But when they called, or returned, they only had one number to call and one place to return, their "Posko Lambung"

All decisions on public affairs are made at the Posko based on *mufakat*. The *gampöng* officers were also put back together as soon as they had first *rapat (mufakat) gampöng* when they were still in the IDP camp. Therefore, the survivors from Lambung reorganized themselves faster and better than, for example, those survivors from Punge Jurong's neighbourhood of Al-Mukarramah. "I noticed the weakness of other villages is that they opened several "poskos" to get the most from relief supplies, but it caused problems of coordination" observed Hardi, the young community organizer from Lambung.

Like many other communities, early returnees in Lambung were all male survivors. Female survivors stayed in host families and did

---

humanized by the crowded barracks and sub-standard service provision at another point.

[42] *Gotong royong* is working together for public needs, a social volunteerism, a community self-help.

[43] Posko stands for "pos komando", or "command post", a place to get together and where organizers work from, a term borrowed from military lexicon.

not return until two more barrack units were erected. Once temporary shelters in the village were available, however, women chose to leave the host communities or other type of temporary shelters and return to the village where they found more acquaintances[44]. When asked what made Lambung survivors re-organize faster and stick together along the way to recovery, Hardi said:

> Sticking together with people of acquaintances who underwent the same experiences ease your burden and sadness. But leadership from our keuchik and sekdes (waki) was also essential in keeping us together. Our keuchik is an achiever (pendobrak) while our waki is a careful planner. But I think we have been very cohesive (kompak) since before the tsunami.

Sadly, however, Hardi observed that the cohesiveness had been eroding after the emergency period, especially when the infamous "cash for work program" were introduced in Lambung.

> Before, many friends of my age and other fellow villagers were willing to work voluntarily, for free, for our own gampöng. Now, there is still some willingness, but I felt that it has been gradually decreased. People started to be lazy to work together, gotong royong, maybe, since the introduction of 'cash for work'. I think, it would be better if they just give the money away, without attaching it to any scheme like "cash for work". No reason needed to "work", cleaning up your own gampöng. It should have been clear difference: donation or earned-money. Don't mix them up. Why they have to make up reasons to give away donation money?

---

[44] One of the conclusions, R. Cibulskis, "Internally Displaced People" (Banda Aceh: World Bank, 2006) accessed on 15 November 2006 at http://www.humanitarianinfo.org/sumatra/reliefrecovery/livelihood/docs/doc/inforesources/IDPsSituationbyRichardCibulskisWB.ppt, had on SPAN data is that "Women more likely to move out from temporary homes."

It is worth noting the use of "they" by Hardi in the interview. It seemed that he used "they" instead of, for example, "NGOs" or more specifically "international NGOs" when he wanted to express his disagreement. "They" is a stronger expression of the otherness for Acehnese.

After Displacement: Finding Back *Gampöng*

Going back to one's own land seemed not to be an individual choice for many tsunami (and conflict) survivors in Aceh. Even when a survivor had been provided with a house, the decision to return might have not been made until at least a number of people from the same original community, *gampöng*, returned. Re-organized communities, in turn, might decide whether to return or not, and if so when, based on other supporting variables like infrastructure (electricity, water and sanitation), livelihoods, and social life. Women and children might not feel safe to return when inhabitants of a settlement are still sparse.

For a settlement to be a settlement for the Acehnese, a *gampong* both in physical and non-physical notions must be rebuilt; not just a collective or rows of houses without a thread to one another, no matter how neat they seemed. People returned to their *gampong*, not merely to their houses or their settlement.

Membership and loyalty to a *gampong*, besides to a family, is most important for Acehnese in coping with hardship after the tsunami. Social structure beyond these two layers is likely to be secondary. Member of civil service institutions in Aceh, which collapsed during the conflict, did not indicate significant loyalties to their post at the wake of the quake and tsunami. This observation is in line with other observations on role conflict in critical situations studied by Lewis Killian[45]. With few exceptions—those who were single and community-oriented and whose selfless acts made them community heroes, Killian documented that most people in disaster settings try to help

---

[45] Lewis M Killian, "The Significance of Multiple-Group Membership in Disaster," *The American Journal of Sociology* 57, no. 4 (1952): 309-314.

their family first. In the case of the 2005 Hurricane Katrina in the Gulf region of southern regions in the US, for example, "Police officers simply walking away from their jobs, in many cases prioritizing the needs of their own families."[46] In the case of Aceh in general, and *Gampöng* Lambung in particular, family, more often than not, also meant a community in a *gampöng* as extended families usually lived in the same *gampöng*. Hence, helping your family was helping your *gampöng* and *vice versa*.

Early returnees from both Al-Mukarramah neighbourhood and Lambung decided to return to their original settlement collectively, albeit gradually, from around mid to the end of 2005. Villagers from Lambung, however, returned in a more organized manner as they had been well organized since they were in displacement. The fact that none of the houses and buildings in Lambung was inhabitable made the people stick together. They had to build a new common building for everyone, initially all male, to stay.

On the other hand, some members of Al-Mukarramah neighbourhood in Punge Jurong had their houses intact after the tsunami. In the first weeks, they did stick together while cleaning and then utilizing the mosque as their community centre and "posko". Most activities, especially in dealing with outside actors, were done in the mosques. Then, early returnees started to access and clear the remains of their houses and properties in the neighbourhood. This had made for a different return experience for members of this community. That is, some stayed at their renovated houses, some stayed on in make-shift structures on their land, others stayed in tents erected on their property. And yet some others temporarily stayed at the mosque *cum* community centre.

As the two communities were returning to their original settlement in Lambung and Punge Jurong, the "common rivals" of "other

---

[46] Chester Hartman, and Gregory D. Squires, "Pre Katrina, Post Katrina," in *There is No Such Thing as a Natural Disaster: Race, Class, and Hurricane Katrina*, eds. Hartman, Chester and Gregory D. Squires (New York: Routledge, 2006), p. 16.

communities" were gone. Tensions among the survivors from one community, within their own village, started to emerge. Tsunami survivors of a village who used to be more united — be it due to solidarity coming out of the nature of the "democratic disaster", or merely due to the fact that they are acquaintances to one another previously; became less cohesive. Those who once are friends and acquaintances, in some cases, could become "others" if not "outsiders", even adversaries. This is apparent even in Lambung which is considered as having a more cohesive community.

It was as if social cohesion was stronger during the emergency period following the tsunami while the two communities were still living outside their villages in their respective temporary shelters. There were tensions as a result of rivalries over aid supplies, but the tensions were more with "outside" and "other" communities. In a way, there seemed to be an understanding that "our village" was "fighting" the "other village" over aid supplies. In addition, in defining the "togetherness" and "membership" of a community, leadership played a pivotal role.

Leadership and governance during displacement influenced the process of community reorganisation once communities decided to return. Al-Mukarramah neighbourhood lost its community chief to the tsunami, while Lambung's head survived. This had significantly impacted the performance of both communities, especially after the emergency period and after the return to their respective village. Although leadership in both communities were somewhat negatively corrupted by the projects and programs from outside donors, leadership in Al-Mukarramah was more damaged by the questions of legitimacy. The fact that Abubakar Ishak, the neighbourhood head of Al-Mukarramah, was elected during the emergency period was considered temporary by some people. That was another reason for Mr. Ishak to withdraw himself from the leadership in May 2006. But when the neighbourhood held an open election in June 2006, Mr. Ishak won the re-election with a landslide result.

Legitimate leadership, however, does not guarantee good governance. In the case of both Lambung and Al-Mukarramah, the leadership was somehow disoriented during recovery and reconstruction process. In contrast to focused and strong leadership during the emergency period, both communities witnessed a presumably weaker leadership when their leaders played the role as local contractors for various programs and projects of outside donors. Although some NGOs and UN agencies involved in the reconstruction projects in the two communities tried to be as participatory as possible, there was an indication of "local elite capture" in various reconstruction projects. The case was more apparent in Al-Mukarramah neighbourhood which has a weaker leadership after the emergency.

The leadership in both communities defended their involvement as contractor or proxy contractor as people wanted the project to maximally benefit the communities. As in many other communities, people did express their opinion that all projects should be contracted to "local" contractors. In many cases, however, including those in the two communities, the local contractors were in fact the few elite groups within the communities, including the leadership. This had caused tension to some extent in both communities. Two persons interviewed in Lambung, for example, were closer friends before one viewed the other as "*punya proyek*" or "having projects". The one who ran the project was actually "chosen" by the villagers themselves. Yet, the projects have alienated some people from others. Does it mean that the "participatory approach" failed?

Defining "participatory" itself is not easy in the context of post-tsunami and post-conflict Aceh. "If you wait until everybody returns to the village, you will never do anything" said one community organizer in Lambung. "Worse, households with no survivors in the village are more difficult to hope for their involvement in the process. They do not live in the village, but wanting every right other survivors in the village get. Participatory is not an easy task!"[47] continued the community organizer.

The village of Lambung—given that most of its inhabitants are native and deeply rooted to the village, with a relatively stronger leadership, seemed to have made social ties stronger in this village, resulting in easier and faster progression in re-organising and reaching a consensus after the tsunami. The village, for instance, reached consensus for land consolidation relatively easily.

Figure 2: Arranged Urban Settlement after Land Consolidation in Lambung

Gampöng Lambung's success in participatory land consolidation has been applauded by local government of Banda Aceh and BRR (Government body for Aceh and Nias rehabilitation and reconstruction). *Gampöng* Lambung has been chosen to be a model village for reconstruction. Seventy billion IDR, about US$7.7 million, had been

---

[47] Interview with Mr. Masykur, 38 years, a community organizer in Lambung.

earmarked by the BRR to develop 42 roadblocks in the gampöng with drainage system, telephone network, public and social facilities.[48] Roads in the *gampöng* are at least six meters wide; some are even 15 meter wide, thanks to the willingness of its people to voluntarily let go of part of their land or consolidate with another fellow villager. When this report was written, the *gampöng* had been visited by many to learn from their experience of land consolidation and village planning, or just to enjoy the "moder village infrastructure it exemplified.

Land certification and consolidation were particularly successful in this *gampöng*, compared to that of the Al-Mukarramah in Punge Jurong or even to those of other villages in Aceh. Both Lambung and Punge Jurong got support for land certification efforts from the Reconstruction of Aceh Land Administration System (RALAS). RALAS supported the National Land Agency (BPN) with US$28.5 million grant to "identify land ownership through a community-based adjudication process and issue land titles to up to 600,000 land owners."[49] The adjudication process was supposed to start with community mapping as an "important prerequisite for the reconstruction of settlements." This process had been completed in Gampöng Lambung, while the same process was not so successful in Punge Jurong. While there were problems of bureaucracy in the BPN

As of December 2007, the two villages seemed to have been in the final state of reconstruction. Thanks to successful "land consolidation", Lambung has been the model village in Banda Aceh. Roads were wide and neat, mostly straight and with no dead end. Open drainage system follows on the side of the roads. Though not very satisfying in terms of building quality, houses were neatly built in rows within

---

[48] "Untuk Bangun Desa Lambung Butuh Dana Rp 70 M," *Serambi Indonesia Daily*, 15 October 2006.

[49] Information on this project is available on the World Bank website: See, "Reconstruction of Aceh Land Administration System (RALAS)," accessed on 30 March 2007 at http://web.worldbank.org/WBSITE/EXTERNAL/ COUNTRIES/EASTASIAPACIFICEXT/INDONESIAEXTN/0,,contentMDK:208 77372~pagePK:141137~piPK:141127~theSitePK:226309,00.html .

Figure 3: Densely Populated Neighbourhood in Punge Jurong (Without Land Consolidation)

blocks. This might be the most arranged urban settlement to date in Banda Aceh (See Figure 2).

The Al-Mukarramah neighbourhood in Punge Jurong seemed to have better houses although the houses had just been finished lately, three years after tsunami, much later than house reconstruction in other villages. This fits into anecdotal belief in Aceh these days: "the longer you wait for housing, the more patient you are, the better houses you will get". It is indeed true in many cases. Not only are new houses, obviously, newer, but also seemed to be better in design, material, and size (Type 42 sq meters floor in the Al-Mukarramah neighbourhood, compared to Type 36 in the other four neighbourhoods in Punge Jurong and in Lambung as well as the same type mostly in other places). However, the neighbourhood, being unable to do land consolidation, was still with a no-grid-system of roads and houses. The roads were wider here and there, but mostly not as straight, manoeuvring among densely built new houses. In some cases, roads and drainage lines were still blocked by fences of remaining houses spared by the tsunami (See Figure 3).

On the new physical outlook of the "modern urban settlement" showcased in Lambung, Sylvia Agustina, a native of Lambung and an urban planner herself, with a graduate training in the US, worried that people might get more individualized by the fact that people in her Lambung village do not need to interact with one another as much as before.

> ....before, as the roads follow the footage of the property; people often have to pass other's property, thus necessitate interaction among neighbours. Now, with straight roads and grid-based houses, people can avoid meeting their neighbours all together and directly access the main road.[50]

While Sylvia is very happy with the re-arrangement of her village as a "neat space", she was worried that the "space" was not the same "place" as she had experienced since her childhood. She, therefore, had been urging that more "public spaces", something not appropriately addressed in Lambung during its physical reconstruction, to be prioritized and quickly built.

> There is an urgent need to bring people back together again as a gampöng; and this requires public spaces where people can meet, children can play, and housewives can socialize; otherwise Lambung will merely be another soul-less settlement, a mere space, not a place[51]

As of December 2007, Lambung was indeed in need of a public space as their mosque *cum* community centre was not yet finished.

---

[50] Author's interview with Sylvia Agustina, Banda Aceh, 2 December 2007.
[51] Author's interview with Sylvia Agustina, Banda Aceh, 2 December 2007.

When asked about public space and activity, Aldi (not a real name), 39 years, one of Lambung native villagers said

> Look at the wide and great quality roads. We don't really need all of these, but we need better quality of houses, and we need jobs. We play volley ball on the wide roads as we do not have empty field and public space as before the tsunami.

Concluding Remarks

Social cohesiveness in the two communities under study changed all the time as the communities spread out after the disaster, partly reunited in temporary shelter, spread out again by the limited provision of barracks (in the case of Al-Mukarramah neighbourhood), and reunited again at their original settlement after they returned. The dynamics in the two communities during displacement, early stage of their return to their original spaces, and consolidation as time goes by, depended mainly on community cohesiveness and the kind of leadership they had before the tsunami. In turn, these influence the dynamic interplay with outside interventions. Lambung and Al-Mukarramah, therefore, had taken different routes to a total recovery. Tensions within Al-Mukarramah community seemed to be higher as the result of geographic divisions such as when some of the community members were in barracks while others had returned to the village, thus creating *a split community*. This was not the case with the Lambung community which remained united until the provision of houses was begun.

In relation to the provision of houses, a "participatory approach" seemed to have been more successful in Lambung. The donor for housing (REKOMPAK/P2KP) funded by the World Bank gave the Lambung villagers full authority to plan and build their houses. They did not use contractors for the housing project; some even built their houses together. At the later stage, however, they did use "local" contractors to expedite the process. Villagers in Al-Mukarramah also demanded that "local" contractors be given the project by their donor, the Indonesian Red Cross with funding from IFRC. This "participatory approach", nevertheless, is not a panacea. From interviews in the two villages, there was tendency that "local" contractors become "the other" when some people viewed them as benefiting more from the process.

It is clear that united communities are better able to negotiate with outside interventions, including international organizations, while others split vis-à-vis solutions offered and thus cannot keep *gampöng* together. Although the two villages are geographically very close together, they differ in other respects: Lambung is socially more cohesive with most inhabitants being native to the village. People in Punge Jurong, particularly of Al-Mukarramah neighbourhood, are less cohesive with most people coming from other regions in Aceh, especially Pidie, as economic and conflict migrants.

Leadership in both communities in the case studies reported here were weakened by suspicion and distrust over aid supplies, but more so over reconstruction projects. The leadership, nevertheless, played an important role in different stages of community re-grouping and activities. Leadership in both communities was pivotal in the early re-grouping of both communities in the temporary shelters. The Al-Mukarramah neighbourhood leadership, however, faced greater challenges when the community was split: some remaining in the barracks while others returned early to their village. On the other hand, leadership in Lambung was united all along the way. When the two communities started the reconstruction of houses, roads, water pipelines, drainage systems, etc, both leadership structures faced greater challenges. There seemed to be "elite capture", at least as perceived by the communities, no matter how "participatory" the approach to programs and projects in both communities were.

Village to village comparison also indicated that the elected village head in Lambung was more effective than the appointed one in Punge Jurong. The village head in Lambung lived and worked along the way with his people, while the village head in Punge Jurong only worked from his temporary office during the day. The bond between the village heads and their people were stronger in Lambung than that of Punge Jurong. That is why, other neighbourhood heads like

Abubakar Ishak of Al-Mukarramah in Punge Jurong were more or less left alone to work with its own community. Although he would need a final approval from the village head, the neighbourhood chief took more initiatives and independent actions. The fact that Mr. Ishak was elected more formally after they returned to their neighbourhood, combined with village head's approval for formal paper works, had helped him to work as effectively as, if not more than, the village head of Lambung.

Are neat, wide and straight roads in a village always good? Interviews with community members in Lambung indicated that it was not always the case. "Why do they build such fancy, wide roads? We don't live on the road," said one interviewee. Another interviewee worried that direct access to a wide road could reduce community members' interactions as people can directly access the road from their houses without having to interact with their neighbours. The notion of "modern" urban setting with neat rows of roads and houses does not always give a sense of community for inhabitants of a "space". Sense of community and interconnectedness among its members is the soul that makes a "space" a "place" to live.

Saiful Mahdi is a lecturer in Statistics at Syiah Kuala University, Banda Aceh, Indonesia. He is also the co-founder and first executive director of the Aceh Institute, a research and policy studies centre based in Banda Aceh.

A Double Wounding?
Aid and Activism in Post-Tsunami Sri Lanka[1]

Malathi de Alwis

Abstract

This paper explores the intertwined notions of charity and wounding in war-torn, post-tsunami Sri Lanka where aid, in all its myriad forms from charity to cash-for-work to psychosocial therapy, appears, at least at first pass, as a salve for those wounded by war and/or the tsunami. The operations of such a 'salve' is further unravelled through a discussion of some key categories mobilised in development/humanitarian discourses and practices such as 'capacity building', 'participation', 'empowerment' and 'good governance.' The paper concludes by addressing what becomes of the 'political' in a context of such

[1] Special thanks to my research assistants Samanmali Kumari Jayawardena, Prabath Hemantha Kumara and Neel Dharmaretnam, to Priyantha Kaluarachchi and Manoja Dharmahewa for sharing the documentary film *At Peraliya* as well as their experiences of making this film and to Edward Simpson and Pradeep Jeganathan for the photographs reproduced here. I am also grateful for the useful comments, questions and suggestions I received when presenting an earlier version of this paper in Adelaide, Colombo, Galway, Los Angeles, Mayneuth and New Delhi. I wish to thank in particular, Sunil Bastian, Sarah Clancy, Akhil Gupta, Jane Haggis, Eva-Lotta Hedman, Pradeep Jeganathan, Purnima Mankekar, Chandana Mathur, Saloni Mathur, Harish Naraindas, Michael Roberts and Jamie Saris. Names of places and individuals have not been changed in accordance with anthropological conventions due to them already being part of the public domain.

profound wounding and in a context where those who are wounded are aided and ad(minister)ed and disciplined in additionally wounding ways.

In her foreword to Marcel Mauss' celebrated text, *The Gift*, Mary Douglas makes the provocative comment that charity wounds.[2] This is a particularly interesting turn of phrase given the unexpected, even unanticipated juxtaposition of charity with wounding. In this paper, I wish to further explore this concept of 'wounding' in war-torn, post-tsunami Sri Lanka where aid, in all its myriad forms from charity to cash-for-work to psycho-social therapy, appears, at least at first pass, as a salve for those wounded by war and/or the tsunami. Such a formulation also begs the question, what becomes of the political in a context of such profound wounding and in a context where those who are wounded are aided and ad(minister)ed and disciplined in additionally wounding ways? In other words, is the reiteration of development aid, now enfolded within humanitarianism, the new 'prose of counter-insurgency'[3] ?

Douglas' ruminations on the gift begins with the argument that the idea of charity or the 'free gift' while "lauded as a Christian virtue" is nevertheless based on a misunderstanding: "The recipient does not like the giver, however cheerful he may be."[4] Note here the affectual rendering of the giver's disposition as 'cheerful' and his gendering; both positionalities are marked as inadequate here. Marianne Gronemayer's thoughtful exposition on 'help' similarly renders this relationship as that of one between a shamed receiver and an arrogant giver.[5] Gronemeyer invokes Thoreau to offer us yet another paradox -- help as a

---

[2] Mary Douglas, "Foreword: No Free Gifts," in Marcel Mauss, *The Gift*, trans. W. D. Halls (New York/London: Routledge, 2002[1954]), p. ix.

[3] Ranajit Guha, "The Prose of Counter-Insurgency," in *Subaltern Studies II*, ed., Ranajit Guha (Delhi: Oxford University Press, 1983).

[4] Douglas, *op. cit.*

[5] Marianne Gronemeyer, "Helping," in *The Development Dictionary*, ed., Wolfgang Sachs (London: Zed Books, 1999), p. 65.

threat, a precursor to danger: "If I knew that a man was coming to my house with the conscious design of doing me good, I should run for my life ... for fear that I should get some of his good done to me."[6]

The pure or free gift wounds because it denies reciprocity and thus leaves the receiver in a position of indebtedness and powerlessness, and over the long term, even of dependence.[7] Present day development workers however, in seeking to circumvent such charges of charity or gifting have resorted to what Roderick Stirrat and Heiko Henkel refer as the 'development gift' -- the practice of "helping the poor to help themselves"[8] or to use a popular buzz word now ubiquitous within this field: 'Capacity building'. The ultimate goal of the development gift thus is 'self realization' and it is through this sleight of hand, as it were, that "help apparently rediscovers its innocence."[9]

The attainment of self-realization -- whether the 'beneficiaries'[10] may wish it or not -- is charted through a vast and arduous terrain signposted with other developmental buzz words such as 'participation', training of trainers, 'gender sensitization', 'gender mainstreaming', 'conflict transformation', 'good governance', and 'empowerment'. I do not wish to rehearse here the excellent genealogy of participation offered by Henkel and Stirrat but do wish to call our attention to its origins in Christian traditions imbued with evangelical promises of salvation

---

[6] Quoted in Gronemeyer, *op. cit.*, p. 53

[7] Douglas, *op. cit.*, p ix. See also Roderick Stirrat and Heiko Henkel, "The Development Gift: The Problem of Reciprocity in the NGO World," in *Annals of the American Academy of Political and Social Science* 554 (1997): 66-80.

[8] Stirrat and Henkel, *op. cit.*, p. 73.

[9] Gronemeyer, *op. cit.*, p. 66.

[10] The continued use of an asymmetrical signifier such as 'beneficiary' even within discourses framed in the terms of 'self-help'/'self realization'/'capacity building' highlights the contradictions which underpin this way of thinking. For a problematization of this label, see Barbara Harrell-Bond, "Can Humanitarian Work with Refugees be Humane?," *Human Rights Quarterly* 24, no. 1 (2002): 51-85 and Jennifer Hyndman and Malathi de Alwis, "Beyond Gender: Towards a Feminist Analysis of Humanitarianism and Development in Sri Lanka," *Women's Studies Quarterly* XXXI, nos. 3 & 4 (2003): 212-226.

through the "participation of man in the infinite grace of God" -- i.e., in the reading of scriptures, in governing the church and most importantly, "actively participating in the duties of the community."[11] It is thus a "culturally specific concept", note Henkel and Stirrat, rather than a "matter of universal common sense."[12]

By baring the old and powerful religious heritage of what at first pass seems a profoundly secular concept, especially given its better known historic roots in bourgeois emancipation in 18th and 19th century Europe, Henkel and Stirrat also illuminate the "missionary habitus"[13] of the guru of this new "theology of development,"[14] Robert Chambers,[15] who has single handedly made tremendous strides in the promotion of now extremely popular participatory mechanisms such as PLA (Participatory Planning and Action), PID (Participatory Interaction in Development), RRA (Rapid Rural Appraisal), and especially, PRA (Participatory Rural Appraisal):

> The PRA approach centres [on] an institutionalized, communal construction of visual representations of reality, i.e. maps, matrixes, diagrams and calendars. PRA, so its proponents often claim, leads to the 'empowerment' of local people by giving them both the means to analyse the problems of their communities

---

[11] Heiko Henkel and Roderick Stirrat, "Participation as Spiritual Duty; Empowerment as Secular Subjection," in *Participation: The New Tyranny*, eds., Bill Cooke and Uma Kothari (London: Zed Books, 2001), pp. 173-4. For a brief genealogy of the shift from saving souls to social regimentation, see Gronemeyer, *op. cit.*

[12] Henkel and Stirrat, *op cit.*, p. 175.

[13] *Ibid*, p. 178.

[14] *Ibid*, p. 177.

[15] See Robert Chambers, *Rural Development: Putting the Last First* (Harlow: Longman, 1983); Robert Chambers, *Rural Appraisal: Rapid, Relaxed and Participatory*, IDS Discussion Paper 311 (Sussex: Institute of Development Studies, 1992); Robert Chambers, *Paradigm Shifts and the Practice of Participatory Development*, IDS Working Papers, no. 2 (Sussex: Institute of Development Studies, 1994); Robert Chambers, *Whose Reality Counts? Putting the Last First* (London: Intermediate Technology, 1997).

on their own terms and a voice vis-à-vis external policy makers such as politicians and development experts.16

The insidious aspect of PRA is that what is presented as supposedly neutral visual techniques in actuality encourages a particular way of seeing, understanding and representing the world, which derives from the world of the PRA 'expert.'17 It is possibly these kinds of subtle interventions which also leads Gronemeyer to observe that 'help' is one of the most elegant exercises of power as "it is unrecognizable, concealed, supremely inconspicuous."18

---

16 Henkel and Stirrat, *op. cit.*, p. 179.

17 *Ibid*, p. 182.

18 Gronemeyer, *op. cit.*, p. 53. An earlier, somewhat more schematic analysis of 'participation' by Stirrat usefully lays out many of the problematic, neo-Orientalist assumptions on which it is based such as idealized notions of 'village', 'community', and 'rural folk' and naïve conceptualizations of power and knowledge (see R. L. Stirrat, "The New Orthodoxy and Old Truths: Participation, Empowerment and Other Buzz Words," in *Assessing Participation: A Debate from South Asia*, eds., Sunil Bastian and Nicola Bastian (New Delhi: Konark, 1996).
For a critique of participatory methodologies through a different lens –i.e., the production and commodification of 'people's knowledge' which is nevertheless subordinated in the discourses and practices of project planners and policy implementers thus emphatically discounting the current development myth that participatory approaches have brought about a paradigm shift in development planning - see David Mosse, "Authority, Gender and Knowledge: Theoretical Reflections on the Practice of Participatory Rural Appraisal," in *Development and Change* 25, no. 3 (1994): 497-525; David Mosse, "The Social Construction of 'People's Knowledge' in Participatory Rural Development," in *Assessing Participation: A Debate from South Asia*, eds., Sunil Bastian and Nicola Bastian (New Delhi: Konark, 1996); David Mosse, "'People's knowledge', Participation and Patronage: Operations and Representations in Rural Development," in *Participation: The New Tyranny*, eds. Bill Cook and Uma Kothari (London: Zed Books, 2001); David Mosse, *Cultivating Development: An Ethnography of Aid Policy and Practice* (New Delhi: Vistaar Publications, 2005).
For a useful mapping of the different ways in which 'participation' is mobilized in the plantation sector in Sri Lanka and how they intersect with political struggles and political participation, see Sunil Bastian, "Living Between 'Participation' and 'Self Determination'- Some Reflections on the Plantation Sector in Sri Lanka," in *Assessing Participation: A Debate from South Asia*, eds. Sunil Bastian and Nicola Bastian (New Delhi: Konark, 1996).

Forms of representation, such as PRA, are productive of a whole new way of thinking and organizing social life: Individuals are encouraged to enter into a form of social contract with each other, during PRA exercises, which elides existing social hierarchies and dependencies (as such public 'performances' are often dominated by the more articulate, elite, male members of the community)[19] and subsequently becomes the foundation for the construction of a 'modern' community.[20] Those who argue for participation in development, notes Bastian, "tend to believe that through this notion they have found a way of dealing with social contradictions generated by capitalism."[21] Indeed, participation has "replaced many other policy prescriptions for dealing with problems of the disadvantaged, such as land reform, expansion of investments by the state etc."[22] Even more insidiously, the presentation of PRA initiators, 'experts' and animators as mere 'facilitators' in this process enables the development industry to shift responsibility (and thus accountability) for the consequences of their projects away from themselves and to the participating volk.[23]

The articulation of 'empowerment', within such a scenario, is equally problematic. It is based on an *a priori* assumption that 'the other' is powerless despite an extensive, existent literature on a variety of forms of resistance to/negotiation of domination which are part and parcel of everyday rural life.[24] Further, this paradoxical move of 'disempowering' in order to 'empower' produces yet another paradox, albeit a seemingly morally legitimized one: the creation of new dependencies.[25] By disrupting the capacity of a community to shape and maintain its way of life through interventions to promote the autono-

---

[19] See Mosse 1994, *op. cit.*

[20] Henkel and Stirrat, *op. cit.*, p. 182.

[21] Bastian, *op. cit.*, p. 243.

[22]*Ibid*, p. 243.

[23] Henkel and Stirrat, *op. cit.*, p. 183.

[24] For some citations, see R. L. Stirrat, *op. cit.*, fn 23; Bastian, *op. cit.*, p. 251.

[25] R. L. Stirrat, *op. cit.*, p. 76; Gronemeyer, *op. cit.*, p. 66.

my of the oppressed, exploited and marginalized such as women, widows, lower castes and classes, ethnic and sexual minorities etc., a new elite of 'motivators' and 'mobilisers' on whom the newly 'empowered' now become dependent, is spawned.[26] Even more insidiously, 'empowerment' through 'participation' becomes "not just a matter of 'giving power' to formerly disempowered people" but also of incorporating them in the "great project of 'the modern'" -- as responsible citizens, as diligent consumers, as rational farmers etc.[27] This attempt to "reshape the personhood of the participants", further note Henkel and Stirrat, is tantamount to what Foucault calls 'subjection.'[28] Participation, however counter-intuitive it may seem, thus transforms into a form of governance.[29]

A consideration of this form of governance, extrapolated through micro techniques of power such as PRA, becomes especially productive when juxtaposed with Rob Jenkins' concise yet illuminating formulation that aid agencies mistake 'governance' for 'politics.'[30] Jenkins' category of analysis is civil society,[31] another popular buzz word

---

[26] R. L. Stirrat, *op. cit.*, p. 76.

[27] Henkel and Stirrat, *op. cit.*, p. 183.

[28] Henkel and Stirrat, *op. cit.*, p. 182; Michel Foucault, "Afterword: The Subject and Power," in *Michel Foucault: Beyond Structuralism and Hermeneutics*, eds., Humbert L. Dreyfus and Paul Rabinow (Chicago: University of Chicago Press, 1982).

[29] Henkel and Stirrat, *op. cit.*, p. 178.

[30] Robert Jenkins, "Mistaking 'governance' for 'politics': foreign aid, democracy, and the construction of civil society," in *Civil Society: History and Possibilities*, eds., Sudipta Kaviraj and Sunil Khilnani (Cambridge: Cambridge University Press, 2001). *Cf.*, Firoze Manji and Chris O'Coill, "The Missionary Position: NGOs and development in Africa," in *International Affairs* 78, no. 3 (2002): 567-83, who argue that the "'good governance' agenda" and the "co-opting" of NGOs and other civil society organizations, in Africa, in the 1990s, sought to provide a "human face" to neo-liberal economic and social programmes which had previously been met with protests, strikes and riots, in the 1980s. See also G. Cornia, G. Jolly, and F. Stewart, *Adjustment with a Human Face* (Oxford: Clarendon Press, 1987).

[31] In this regard, he is one of many scholars who have sought to critique the concept of civil society which is mobilized in post colonial societies. For example, Mahmoud Mamdani has argued that civil society in Africa worked in collusion with their white

within development circles, and a community to which even humanitarian aid organizations often turn to in Sri Lanka, on the premise that humanitarian interventions must go hand in hand with 'conflict transformation', i.e., finding a political solution to the ethnic conflict in the country. Such a turn, involves the deploying of considerable economic and political resources to such organizations as well as attempts to influence, more broadly, the particular political milieu in which they must operate. The Norwegian government seeking to facilitate a ceasefire in Sri Lanka while also financially supporting many peace groups in the country, is a case in point.

Jenkins, persuasively sets out a variety of contradictions which inhere in donor governments' support of civil society organizations in post colonial societies and concludes that it has led to them being depoliticized, sacralized and bureaucratized: the formation of opposition movements, the exercise of state power etc., are grossly underplayed while civil society emerges as a sort of "political ombudsman, reflecting the values of impartiality, fair play, and commitment to public welfare."[32] In fact, this kind of value orientation is very similar to the one

---

colonizers and continues to sustain neo-colonial hierarchies today [see Mahmoud Mamdani, *Citizen and Subject: Contemporary Africa and the Legacy of Late Colonialism* (New Jersey: Princeton University Press, 1996)], while Partha Chatterjee returns to the old idea (i.e., of Hegel and Marx) of civil society as bourgeois society to nevertheless address its working out within a different form of modernity i.e., one encountered through colonialism (Partha Chatterjee, "Beyond the Nation? or Within?," in *Social Text* 16, no. 3 (1998): 57-69; Partha Chatterjee, "On Civil and Political Society in Post-colonial Democracies," in *Civil Society: History and Possibilities*, eds., Sudipta Kaviraj and Sunil Khilnani (Cambridge: University of Cambridge Press, 2001); Partha Chatterjee, *The Politics of the Governed: Reflections on Popular Politics in Most of the World* (Delhi: Permanent Black, 2004). This civil society of citizens, shaped by the normative ideals of western modernity is only a small, elite group which assumes a 'pedagogical mission' of enlightenment towards the vast, excluded mass of the population which Chatterjee places within 'political society'.

[32] Jenkins, *op. cit.*, p. 268. Activist Aziz Choudhry provides a more trenchant commentary when he notes that civil society craftily steers grassroots organizations away from political organizations calling for radical and comprehensive political reforms (Aziz Choudhry, "All this 'Civil Society' Talk Takes us Nowhere," in *Znet*, 9 January 2002). Similar sentiments are echoed by Nepali Maoists who have accused civil society activ-

that donor agencies see themselves as occupying in relation to the countries to which they give aid. "[I]t should not surprise us after all", notes Jenkins, "that aid agencies have created civil society in their own image."[33]

These unravellings of aid interventions, however truncated in their presentation here, enables me to raise a more crucial question regarding what constitutes radical political struggles within such a landscape; that is to say, struggles which are not articulated in the terms or agendas of aid agencies nor struggles which have been instigated/usurped by nationalisms (extending in some cases to armed struggles) or political parties. I have extrapolated elsewhere on the notion of the 'political' which informs my inquiry here[34] but let me reprise it briefly here again: It is primarily an engagement with Judith Butler's re-working of the 'constitutive outside' (*pace* Derrida) of political fields as laid out in the influential arguments of political theorists such as Ernasto Laclau and Chantal Mouffe.[35] Butler seeks to problematise their insistence that the political field is constructed through the production of a "determining exterior" which leads to the naturalization of a "pre-" or "non" political by arguing instead for a political field which "produces and *renders contingent* the specific parameters of that

---

ists of "working as imperialist stooges to divert the rural masses from real contradictions and struggles (Saubhagya Shah, "From Evil State to Civil Society," in *Himal*, November 2002, p. 2 accessed on 28 March 2008 at http://www.himalmag.com/2002/november/essay/htm).

[33] Jenkins, *op. cit.*, p. 268.

[34] Malathi de Alwis, "Feminism," in *A Companion to the Anthropology of Politics*, eds., Joan Vincent and David Nugent (Boston: Blackwell, 2004); Malathi de Alwis, "Interrogating the 'Political': Feminist Peace Activism in Sri Lanka," in *Feminist Review* 91 (in press).

[35] Judith Butler, "Contingent Foundations: Feminism and the Question of 'Postmodernism'," in *Feminists Theorize the Political*, eds., Judith Butler and Joan Scott (New York & London: Routledge, 1992); Ernesto Laclau and Chantal Mouffe, *Hegemony and Socialist Strategy: Towards a Radical Democratic Politics* (London: Verso, 1985).

constitutive outside."[36] This re-formulation provides for a constant elaboration of differential relations or mutual antagonisms which puts the parameters of the political itself into question.

Indeed, several months after the tsunami, there were many collective, local responses from affected communities which one could argue were both contingent and oppositional however limited their demands might have been. Such responses were triggered by the unilateral institution of a buffer zone by the Sri Lankan state, post-tsunami, resulting in the banning of all reconstruction as well as new construction up to 200 metres from the shoreline, in all tsunami affected areas.[37] The responses took many forms and significantly varied according to region and ethnic grouping. While there was widespread dismay expressed by those affected by this buffer zone ruling, the most significant agitations took place in certain areas in the south which were predominantly Sinhala, and certain areas in the east which were predominantly Muslim.[38]

These pockets of resistance in themselves illuminate prior contourings of the Sri Lankan polity: A Tamil minority which has been cowed into submission by a prolonged war where they have been at the brutal butt end of both the Tamil militants and the Sri Lankan state; they have learnt at great cost that the LTTE does not tolerate dissent, that the government of Sri Lanka perceives them all as collaborators, and that both parties to the conflict are indifferent to their wants and sufferings and are quick to use them as pawns whenever the opportunity arises. A Muslim minority which is often caught in the crossfire between Tamil militancy and Sinhala hegemony but who have tenaciously struggled against both and are now well represented in the

---

[36] Butler 1992, op. cit., fn 2.

[37] The LTTE imposed a buffer zone of 300 metres in the areas they control in the northern and north eastern regions of the country.

[38] A thought provoking anomaly however was the defiant and courageous stand taken by Tamil fishermen in Point Pedro, Jaffna who refused to be cowed by the buffer zone rulings of both the LTTE and the government of Sri Lanka (*Sunday Times*, 3 July 2005).

government as well as the Opposition by different factions of the powerful Sri Lanka Muslim Congress. And a Sinhala majority with a
strong sense of entitlement that the predominantly Sinhala state is
there to serve them, and with a history of rebellion whenever this contract is considered to have been broken.

One of the central concerns of those whose homes fell within
the buffer zone was that while the majority of them -- particularly
those who had small children -- were willing to relocate to higher
ground or the interior of the island (especially after several new tsunami scares, and in spite of the fact that it would make the plying of livelihoods such as fishing near impossible), they were extremely concerned
that the 'tsunami goldrush' of foreign aid would pass them by as the
government was delaying to allocate new land where their new houses
could be built. Thus, many of their agitations took the form of picket-
ing outside local government offices, taking over government offices for
several days, booing and stoning politicians, commandeering state as
well as private lands, erecting billboards on the roadside to publicise
their concerns, performing religious ceremonies calling for revenge on a
tardy government, noisy demonstrations and public meetings, disrup-
tion of train services by sitting on rail tracks, and the obstruction of key
coastal roads both in the south and east.

None of these strategies were particularly new or sustained over
a long period but they were nevertheless able to garner much publicity
given their additional newsworthiness as 'tsunami follow-up stories.'
Indeed, the forcible occupation of the Kalmunai Divisional Secretariat's
office, for over a week, by approximately 400 irate Muslim families, elicited a 30-minute satellite video conference with President Rajapaksa
no less who promised to build them over 700 houses in three months.[39]
A more unusual strategy adopted by the 'Peraliya 100 Meter Tsunami
Refugees Organization' did not garner a similar response by the President or even local government bureaucrats but succeeded in capturing

---

[39] *Daily Mirror*, 2 March 2006. Spokespersons for this group were savvy enough to
demand that this assurance be given in writing! (*Daily Mirror*, 27 February 2006).

a much wider audience and even the attention of the international media. This group erected a giant billboard beside the Southern coastal road demanding that they be 'saved' from a sure death which awaited them due to this cruel buffer zone policy. The 17 bullet points written in Sinhala and English (the latter a very poor translation of the concise, lyrical, alliterative slogans used in the Sinhala text[40]) variously excoriated the government and opposition parties for being self serving and the media and human rights organizations for their silence: "How is it that only foreigners noticed our pain and responded with kindness? Are our own politicians and bureaucrats blind to our suffering?" or "How about handing us a small portion of all those dollars, pounds and euros which you bureaucrats have acquired for yourselves?"

This organization also staged some well publicized protests along the main coastal road by blocking it with huge logs, fishing boats, bicycles, and their own bodies. Several who were arrested for obstructing the road were quick witted enough to give the names of neighbours and relatives who had died in the tsunami, to the arresting officers. This was of course only possible in a context where the argument that one's national identity card had been washed away in the tsunami, would hold water, as it were. It also involved the collusion of previously fractious community members who were willing to unite against what was in their perception a punitive and uncaring state. Without verifiable identificatory documents and addresses for which there were no longer houses, the court orders which subsequently arrived in the names of the dead, had to be trashed!

These agitations, along with several other factors such as a more populist President being elected, the visit of Bill Clinton to Sri Lanka, and I think, most crucially, the state being hard pressed to find so much build-able land in the interior of the island, is credited with making the government back down on this unilateral ban and the institu-

---

[40] For example: "(those in) Hambantota and Matara are treated like kings, we are treated like dogs" (*hambantotata, matarata raja salakili, apata balu salakili*) is merely translated as: "Hambantota-Matara High regards. But not for us."

tion of a more equitable, variable and viable buffer zone compatible with prior coast conservation regulations.

The anti-buffer zone agitations were primarily initiated by men from the local communities who were directly affected by the no-construction ban. Several of them had honed their political teeth in trade union actions in their workplaces such as the Colombo Port and the Ceylon Transport Board (CTB), as local representatives of underground, political movements as well as mainstream political parties,

**Figure 1: Temporary Memorial (2006), at mass grave,
Peraliya, Southern Province**

and as office bearers of community organizations such as fishermen's cooperatives, welfare societies, and temple and mosque management committees. However, most of these protests were soon usurped by more organized political parties such as the United National Party (in the south) and the Sri Lanka Muslim Congress and its splinter groups (in the east). I would argue that this was due to the fact that there exists very little space in Sri Lanka today for autonomous struggles of this sort; much of this space is either dominated by mainstream political parties or 'civil society' organizations such as micro credit societies and 'women's development' and 'conflict transformation' groups who have been 'empowered' by Non-Governmental Organisations (NGOs) and International Non-Governmental Organisations (INGOs). While the former group uses such local struggles to further their own ends by expanding their coterie of supporters and articulating a critique of the governing party, the latter groups encourage individuals to either become more self-focused and a-political or dependent on 'trainings', 'workshops' and 'plans of action' enabled by various I/NGOs to inspire them. A feature common to both kinds of groups is the inhibiting of independent thought and innovative action.

A welcome shift in such a context, albeit momentary, came in the form of the 'Collective of Those Affected by the Peraliya Train Incident.' As the group's name suggests, it was made up of some survivors as well as family members of those who had perished in the *Samudra Devi* (Ocean Queen) train which was swept off the tracks at Peraliya, during the tsunami. This organization was founded and led by youthful and dynamic Priyantha Jayasiri Perera, a freelance journalist, who had had a lucky escape, along with his mother, but lost his wife of a few months, and his mother-in-law. The central issue which was taken up by this group was to demand that the Sri Lankan state, and the Department of Railways in particular, should take responsibility for the excessive number of deaths which occurred on the *Samudra Devi*. This was based on arguments which ranged from the fact that the train could have been detained at the previous station which was located more inland, or ordered to shunt back to that station, as soon as word reached the Department of Railways regarding the tsunami which had hit the eastern province (there was a gap of at least half an hour before it hit the south-western coast),[41] and that department officials should

---

[41] The Department did manage to stop and turn back some other trains, around this time.

have arrived with helicopters and trained personnel to rescue many of those who were wounded and trapped inside the train or under debris, once the tsunami receded.

The Collective made use of various commemorative rituals beside the mass grave which was dug to bury many of the unidentifiable victims of the train disaster, to publicize their anger and frustration by making speeches at the grave site, standing with placards on the main road abutting the mass grave, and erecting a temporary memorial which stated among other things, that these deaths were not merely due to the tsunami but also to "those in authority neglecting their responsibility" (see Figure 1). This memorial however, has to be constantly re-built as the Collective has been unable to erect a permanent memorial due to the owner of the land on which the mass grave had been dug being engaged in a dispute with the state regarding compensation for unlawfully using his land as a gravesite. To add salt to the Collective's wounds however, the state has circumvented the dispute over land as well as the dispute over responsibility for the train disaster by building its own memorial to the train victims, on the plot of land adjoining the mass grave (see Figure 2)

The most significant intervention of the Collective however was their ability to get the state to appoint a tri-partite Committee to inquire into the culpability of the Department of Railways with regard to the train disaster and the over 1,200 deaths which resulted. A further victory was had when the Committee, comprised of 3 high court judges, who cross-examined over 80 people -- including 20 survivors as well as many Railway Department officials -- produced a verdict which found the Department of Railways culpable and negligent.[42] This verdict, though poorly publicized, nonetheless provided many of the members of the Collective with a feeling that they had been able to

---

[42] See *Report of the Committee to Inquire into the Train Disaster at Telwatte due to the Tsunami of 26th December 2004* (Colombo: Department of Railways, 2005). I am indebted to Prabath Hemantha Kumara who spent several weeks badgering officials in the Department of Railways in order to obtain a copy of this report.

Figure 2: Detail from mural at state tsunami memorial,
Peraliya, Southern Province

somewhat vindicate the deaths of their loved ones by 'seeing justice
done' and hopefully setting in place certain procedures which would
assure that similar disasters could be averted, in the future. Though
several officials associated with the Department of Railways refused to
acknowledge the department's culpability in this disaster, when interviewed in 2008, a news item announcing the launching of an inte-

grated, centrally-controlled signal light system and the distribution of communication sets to train drivers, did quote the Minister of Railways, Mr. Dulles Alahapperuma, as noting that these technological innovations would enable the prevention of "disasters similar to the one that occurred at Telwatta during [the] tsunami in 2004."[43]

What was most noteworthy about this Collective was that it managed to remain autonomous while simultaneously producing a consistent critique of the state's handling of the train disaster. One of the reasons for this is possibly due to the fact that several of its members were from the middle classes and thus much more assertive and influential in their own right and also not dependent on the largesse of humanitarian aid organizations. In addition, this Collective was founded and led by a very politically astute young man whose critique of politicians and society in general was both eloquent and perceptive as is well demonstrated in Priyantha Kaluarachchi's excellent documentary *At Peraliya* (2006) in which he features very prominently.[44] It is unfortunate however, that this Collective's critique was not able to extend beyond the resort to legal measures and the specificities of the Peraliya train disaster.

Why is it that we in Sri Lanka have been unable to sustain a destabilizing and contesting of the 'political' which can also extend to providing more innovative and radical solutions to the ethnic conflict which has been ravaging the island for almost three decades? Have we,

---

[43] *The Island*, 8 July 2008. The Committee's Report notes that all drivers and guards had been provided with walkie talkies, since January 2002, but much of this equipment had stopped being used due to them constantly malfunctioning. (*Report of the Committee to Inquire into the Train Disaster at Telwatte due to the Tsunami of 26th December 2004, op. cit.*, p. 9). Similarly, the centrally controlled signal light system was supposed to have been installed in December 2003 but the Dutch firm contracted for this work, M/s Vialis NMA, had continuously reneged on their deadlines (*Report, op. cit*, pp. 18-19). It is not surprising then that the Department of Railways has done their utmost to 'bury' this report, which is rife with similarly troubling bombshells of negligence and corruption.

[44] Unfortunately, I was unable to meet with Priyantha Jayasiri Perera as he had emigrated to Italy, in 2006.

as political activists, been unable to address a far deeper malaise which is afflicting our society? A malaise akin to that of psychic numbing after several decades of living with unrelenting violence and atrocity upon this 'lacerated terrain,' to borrow a phrase from the poet Jean Arasanayagam?[45] I have suggested elsewhere that perhaps the need of the hour is to interrogate the 'political' via more affectual categories such as grief, injury and suffering.[46] And indeed, the response of the 'Collective of Those Affected by the Peraliya Train Incident', and its founder and leader in particular, has been exemplary in this regard. They have bared their wounds in public in order to produce a critique of the state and effect changes to procedures relating to railway emergencies, which went beyond their individualized injury and grief. Yet, the question remains for me why we are unable to extend our cognisance of this kind of injury, be they ideological or environmental in nature, to address the deepest, darkest and most desolating laceration of our land – the ethnic conflict.

Malathi de Alwis is a Socio-Cultural Anthropologist. She teaches in the Faculty of Graduate Studies, University of Colombo.

---

[45] Jean Arasanayagam, "Numerals," in Jean Arasanayagam, *Reddened Water Flows Clear* (London and Boston: Forest Books, 1991).

[46] Malathi de Alwis, "Tracing absent presence: Political community in the wake of atroci *States of Trauma*, eds., Parama Roy, Piya Chatterjee and Manali Desai (Delhi: Zubaan, in press).

Governmentality, Displacement and Politics: A Witches Brew in Post-Tsunami Aceh, Indonesia[1]

Eva-Lotta E. Hedman

Abstract

This paper explores the emergence and significance of the phenomenon of collective action in the name of the internally displaced in post-tsunami Aceh, Indonesia. To that end, it examines two separate campaigns, each of which involved thousands of people framing local protest in a universalist language of (individual) rights and (government) responsibilities that resonated with the Guiding Principles on Internal Displacement. As instances or 'cases' of entangled encounters between, on the one hand, the novel forms of governmentality introduced to regulate and improve the lives of 'IDPs' in post-tsunami Aceh and, on the other hand, their intended 'beneficiaries' or subject populations, these campaigns help illuminate the ways in which something new is created, through

[1] Research for this paper was carried out during visits to Aceh between February 2005 and December 2007, with the benefit of Aceh Institute as local host institution in Banda Aceh. Research was funded by the International Development Research Council, and also with generous support from the Mellon Foundation and the Refugee Studies Centre, University of Oxford. Lukman Age provided thoughtful interventions and useful introductions in the course of this research. Special thanks also to Saiful Mahdi, Malathi de Alwis, Jennifer Hyndman, and other participants at the Aceh Institute research workshop where an early version of this paper was presented in December 2007.

political contestation and collective action. While Foucault's notion of 'governmentality' is an important and obvious point of departure for exploring the effects of post-tsunami programs of interventions ostensibly aimed at the amelioration of 'IDP' welfare, a closer look at forms of contestation brings into sharper focus political processes, practices and struggles that point beyond such improvement schemes. By focusing on 'IDP protest,' the paper also points to ways of engaging – in theory and method – important questions about the politics of representation and displacement. More generally, it reflects a deeper concern that (the relative silencing of) the voices and experiences of actually existing displaced persons inform forced migration studies, practice and policy.

As documented in numerous reports, surveys, and documentaries, the destruction to life, landscape, livelihood and property left in the wake of the December 2004 tsunami was enormous, and nowhere more so than in the embattled Indonesian province of Aceh, close to the epicenter of the earthquakes that set off such massive waves across the Indian Ocean. After the first few days, the Indonesian government broke the initial deafening silence and called for international support for emergency relief and reconstruction efforts, thus bringing to an end the pre-tsunami virtual lockdown of Aceh, a province where some 40,000 Indonesian troops had been engaged in counter-insurgency campaigns while foreign journalists, human rights observers and other concerned parties remained effectively barred from entry since the declaration of martial law in May 2003.[2] In addition to the unprecedented mobilization of international resources and expertise to assist tsunami survivors across affected areas, Aceh saw a groundswell of local relief efforts, ranging from initiatives by university students and company employees, to missions by religious organizations and political parties. Breaking Jakarta's isolation of Aceh, individual citizens, national governments and foreign militaries, as well as charity, faith-based and other

---

[2] See, for example, Eva-Lotta E. Hedman, ed., *Aceh During Martial Law*, RSC Working Paper, no. 24 (Oxford: University of Oxford, 2005).

non-governmental organizations (NGOs) from the region and beyond, arrived *en masse* to contribute to the relief efforts.[3]

In the years that have passed since the December 2004 tsunami struck, Aceh's political and social landscape has undergone an extraordinary and far-reaching transformation. As the peace-process gained ground in 2005 and the contestation for hearts and minds came to focus on local elections in 2006, the new Aceh witnessed the resurgence of political activism from a wide spectrum of peasant, farmer, worker, student, women and religious groups and movements. While the lives of hundreds of thousands remained, in important respects, defined by their displacement during much of this period, the mobilization of collective action in the name of the internally displaced, inflecting familiar repertoires of protest with a rights-based discourse, also emerged as a distinct phenomenon in this new political context.

Awaiting further scholarly analysis into the nature and direction of wider – political, economic, social, and cultural – change in the new Aceh, this paper explores the emergence and significance of a more specific phenomenon – collective action in the name of the internally displaced. To that end, it examines two separate campaigns, each of which involved thousands of people framing local protest in a universalist language of (individual) rights and (government) responsibilities that resonated with the *Guiding Principles on Internal Displacement*. As instances or 'cases' of entangled encounters between, on the one hand, the novel forms of governmentality introduced to regulate and improve the lives of 'IDPs' (Internally Displaced Persons) in post-tsunami Aceh and, on the other hand, their intended 'beneficiaries' or subject populations, these campaigns help illuminate the ways in which something new is created, or how "a difference is introduced into history in the form of politics."[4]

---

[3] For early glimpses, see, for example, Edward Aspinall, "Indonesia after the Tsunami," *Current History* (March 2005): 105-109.

[4] Nicholas Rose, *Powers of Freedom: Reframing Political Thought* (Cambridge: Cambridge University Press, 1999), p. 26.

Research for this paper has spanned over several field visits and related research between February 2005 and December 2007. It involved the collection of local newspaper articles, and a range of other published and unpublished documents, as well as numerous and sometimes repeated interviews and visits with displaced persons and other concerned parties, including representatives and employees of national and local government, non-governmental and international organizations, as well as journalists, researchers, and many others from a wide cross-section of Acehnese society. Serendipitously, moments of collective action in the name of the internally displaced materialized during the course of research, thus also allowing for an occasional first-hand perspective on the unfolding of a critical challenge in the cogs of the post-tsunami "anti-politics machine."[5]

While Foucault's notion of 'governmentality' is an important and obvious point of departure for exploring the effects of post-tsunami programs of interventions ostensibly aimed at the amelioration of 'IDP' welfare, a closer look at forms of contestation brings into sharper focus political processes, practices and struggles that point beyond such improvement schemes. Recent revisionist scholarship on development in Indonesia has already explored similar dynamics of governmentality and the "practice of politics," and offers further points of reference for the present paper.[6] By focusing on 'IDP protest,' the paper also points to ways of engaging – in theory and method – important questions about the politics of representation and displacement. More generally, it reflects a deeper concern that (the relative silencing of) the voices and experiences of actually existing displaced persons inform forced migration studies, practice and policy.[7]

---

[5] James Ferguson, *The Anti-Politics Machine: "Development," Depoliticization, and Bureaucratic Power in Lesotho* (Minneapolis: University of Minnesota Press, 1994).

[6] Tania Murray Li, *The Will to Improve: Governmentality, Development and the Practice of Politics* (Durham, NC: Duke University Press, 2007).

[7] See, for example, David Turton, *Refugees and Other Forced Migrants* , RSC Working Paper , no. 13 (Oxford: University of Oxford, October 2003).

Governmentality After the Tsunami

In sharp contrast to the enforced isolation that characterized the province during the militarised conflict in the years leading up to the tsunami, Aceh has since attracted unprecedented international attention and resources as the focus of large-scale, multi-donor relief, reconstruction and conflict resolution efforts. In what appeared to be record time for organizational responses to complex humanitarian emergencies, Bappenas *(Badan Perencanaan dan Pembangunan Nasional)*, the Indonesian government national planning board, in collaboration with key international donors, made public two "technical" reports on 19 January 2005 to provide a "Blueprint" that expressed their intention "to start making decisions on setting priorities and considering how to develop a strategy for reconstruction."[8] While the Jakarta government's own so-called "Master Plan" for the rehabilitation and reconstruction of Aceh and other affected areas was not released until April 2005, draft versions of this document, or certain aspects thereof, were in circulation at least as early as February of the same year, thus anticipating, perhaps in crucial ways, its eventual release in April.[9] Indeed, a series of government statements, reports, and initiatives focused on the reconstruction of Aceh stressed the need for continued strict state control of relief efforts, with repeated public calls for centralized coordination and official registration of all concerned actors and activities. This push for a kind of hypercentralization of humanitarian relief and reconstruction efforts also revealed a pervasive concern with control and surveillance in Aceh, where Indonesian civilian and

---

[8] Bappenas (Government of Indonesia), *Indonesia: Preliminary Damage and Loss Assessment/The December 26, 2004 Natural Disaster* (Jakarta: The Consultative Group on Indonesia, January 19-20, 2005); and *Indonesia: Notes on Reconstruction/Assessment/The December 26, 2004 Natural Disaster* (Jakarta: The Consultative Group on Indonesia, 19-20 January 2005).

[9] Bappenas (Government of Indonesia), *Master Plan for the Rehabilitation and Reconstruction of the Regions and Communities of the Province of Nanggroe Aceh Darussalam and the Islands of Nias, Province of North Sumatera* (Jakarta: Government of Indonesia, 12 April 2005).

military officials displayed considerable anxiety about the arrival and activities of international humanitarian organizations, other International NGOs, foreign volunteers, reporters, and militaries.

Such anxieties were also evident in regard to the problematization and representation of displaced persons in the aftermath of the tsunami. From the outset, it emerged that the "labeling" of persons who had been forced to leave their homes would prove a critical issue in post-tsunami Aceh.[10] As discussed below, such labeling reflected pervasive concerns to circumvent or delimit the very category of the (displaced) subject and, thus, the actual scope for (humanitarian) intervention. It also served to point away from the concerns outlined in the *Guiding Principles on Internal Displacement*, in ways that, while not without challenge, nonetheless left enduring legacies for displaced persons in Aceh.

Since their initial publication in 1998, the *Guiding Principles on Internal Displacement* (*Guiding Principles*) have offered an obvious point of reference as they serve to "identify the rights and guarantees relevant to the protection of persons from forced displacement to their protection and assistance during displacement as well as during return or resettlement and reintegration."[11] Widely disseminated by the Office for the Coordination of Humanitarian Affairs (UNOCHA) and other organizations within the UN system, as well as without, the *Guiding*

---

[10] Roger Zetter, "Labelling Refugees: Forming and Transforming a Bureaucratic Identity," *Journal of Refugee Studies* 4, no. 1 (1991):39-62.

[11] "Introduction: Scope and Purpose," *Guiding Principles on Internal Displacement*, available at www.unhcr.ch/html/menu2/b/principles.htm. This text is an excerpt from the document E/CN.4/1998/53/Add.2, dated 11 February 1998 (paragraph 1). Presented to the Commission on Human Rights in 1998 by the then Representative to the UN Secretary-General on Internally Displaced Persons, Francis Deng, the *Guiding Principles* have since been recognized as an "important international framework for the protection of internally displaced persons" at the World Summit in New York in September 2005. See, for example, www.idpguidingprincples.org.

Principles outlines the following "descriptive identification"[12] of internally displaced persons:

> Internally displaced persons are persons or groups of persons who have been forced to flee or to leave their homes or places of habitual residence, in particular as a result of or in order to avoid the effects of armed conflict, situations of generalized violence, violations of human rights or natural or human-made disasters, and who have not crossed an internationally recognized State border.13

Despite the prior circulation of the *Guiding Principles* in Indonesia, including in Indonesian translation, and among government officials,[14] however, the government remained notably reluctant to recognize as internally displaced persons, or 'IDPs,' the hundreds of thousands forced from their homes and unable to return in the aftermath of the tsunami in Aceh. Instead, it often referred to the "homeless," or to the "people who lost their homes because of the tsunami" in public discourse, including in meetings convened by the UNOCHA.

Much as the prompt mobilization of an official discourse on homelessness and the homeless remained notably silent on the very

---

[12] J.P. Lavoyer, "Guiding Principles on Internal Displacement: A Few Comments on the Contribution to International Law," *International Review of the Red Cross* 324 (1998), p. 467.

[13] *Guiding Principles, op. cit.*, paragraph 2.

[14] *Mereka yang Mengungsi: Komik tentang Prinsip-prinsip Panduan Pengungsian Internal* (Jakarta: Baris Baru dan Oxfam GB, 2002). This booklet includes the "Guiding Principles on Internal Displacement," written in Bahasa Indonesia, in an appendix. The cover and the main section of the booklet are in the format of a comic book, featuring fleeing and frightened civilians, glimpses of (disembodied) military boots and weapons, courageous human rights workers, and assorted perpetrators of violence, ranging from masked men with automatic weapons to local thugs engaged in intimidation and worse. In addition to Oxfam, this publication was sponsored by UNOCHA (Indonesia), and Internal Conflict Monitoring Centre (ICMC), and it drew on contributions from a number of locally based human rights organizations, including in Aceh, Papua, Ambon, and Pontianak.

questions of individual rights and government responsibility emphasized in the *Guiding Principles,* the Indonesian government's first plan of action targeting tsunami-affected populations also pointed away from concerns for the needs and protection of internally displaced persons emphasized in international humanitarian and human rights legal and normative frameworks. That is, in the early aftermath of the tsunami, the Indonesian government moved with great speed to announce its relocation program, with particular emphasis on temporary "barracks" rather than the rehabilitation or construction of permanent housing.[15] Having first identified a target number of twenty-four barrack camps that would be constructed to house some 35,000 IDPs in "phase one" of this project, the government promptly revised these figures upwards, to thirty-nine such temporary relocation sites built to accommodate 35,000 households, or some 140,000 displaced persons.[16] The swiftness and ease with which the Indonesian authorities involved UN agencies and NGOs in a "Joint Liaison Unit" to further this project appeared remarkable, not least in the face of considerable apprehension in humanitarian circles regarding the encampment of displaced populations. There was also evidence of concern and even outright opposition to the proposed barracks solution from the outset among affected and other concerned local people.[17] Indeed, against the

---

[15] Eva-Lotta E. Hedman, "Back to the Barracks: *Relokasi Pengungsi* in Post-Tsunami Aceh," *Indonesia* 80 (October 2005), pp. 1-19.

[16] See especially Bakornas PBP, *Bulletin*, no. 32 (25 January 2005) and no. 39 (31 January 2005). See also Bakornas PBP, "Report of Joint Government/United Nations/NGO Rapid Assessment Mission of New Relocation Sites," (January 2005) available at http://www.humanitarianinfo.org/sumatra/reference/assessments/ doc/shelter/JointAssessmentRelocationSite-050405.pdf. Bakornas PBP stands for the National Disaster Management and Refugee Coordinating Board.

[17] See, for example, "Indonesians Wary of Relocation Centers," *Washington Post,* 31 January 2005. At Kreung Sabe, south of Calang on the West Coast, for example, villagers were reportedly "determined to go back and rebuild as soon as possible and ... don't want to go into camps," according to UNHCR Asia-Pacific Bureau director, Janet Lim, cited in "Indonesia: UNHCR Tailoring Shelter Solutions to Meet Aceh

backdrop of counterinsurgency campaigns in Aceh, which had featured large-scale forced displacement of civilian populations, sometimes as a deliberate strategy of war under martial law, it was hardly surprising that, in the immediate aftermath of the tsunami, there was little rush to join in the barracks relocation initiative.[18]

Thus, while the official discourse on displacement did not remain unchallenged, it nonetheless served to put would-be partners in international humanitarian relief and reconstruction on notice that translating their respective "will to improve"[19] into actual programs, as urgently required by organizational mandates and budgets, necessarily involved the demarcation of certain (but not other) kinds of interventions as "intelligible and meaningful"[20] (e.g., a 'barracks' camp). The laying of such a field, in turn, anticipated the leaning into the future of certain kinds of expert or technical interventions (e.g. the mapping and provision of 'watsan,' or water and sanitation). In the same vein, it pointed away from other forms of expertise and interventions less easily "rendered technical,"[21] or, in other words, nonpolitical (e.g., the mapping and provision of protection). In the context of Aceh — an area affected by natural disaster *and* protracted militarized conflict — it also served to distance from humanitarian and political consideration thousands of people whose displacement pre-dated the tsunami.

Needs," *UNHCR Update*, 4 February 2005. For a related report, see "Satlak Aceh Timur tak akui pengungsi Kuala Idi Cut," *AcehKita*, 9 March 2005.

[18] Saiful Mahdi, "IDPs and the Problem of Poverty: Evidence from Aceh Conflict and Tsunami IDP Mobility," Paper presented at the 8th Indonesian Regional Science Association (IRSA) International Conference, 18-19 August 2006, University of Brawijaya, Malang, East Java, Indonesia.

[19] Li, *op. cit.*.

[20] Ferguson, *op. cit.*, p. xiii.

[21] Li adapts the term "rendering technical" from Nicholas Rose and refers to it as shorthand for a set of practices concerned with representing "the domain to be governed ... [and] assembling information about that which is included...." Rose, *op. cit.*, p. 33, cited in Li, *op. cit.*. p. 7.

Such interventions also shaped understandings about the transformation underway in post-tsunami Aceh. In particular, countless needs assessments, mapping surveys, and annual reports were produced by myriad (government/UN) agencies and (international/nongovernmental) organizations in the few years following the tsunami. Unsurprisingly perhaps, the 'problems' thus identified often anticipated 'solutions' in keeping with organizational mandates and practices. Similarly, the recommendations offered have typically focused on meeting programmatic 'targets.' Not unlike other parts of Indonesia, moreover, post-tsunami Aceh also emerged as a busy hub for a kind of transnational cottage industry specializing in the assembly of 'data' on conflict and violence, thus adding additional momentum to the production of certain (but not other) forms of knowledge.[22]

In as much as internationally funded projects continued to draw into their orbit local university lecturers and students, as well as a wide spectrum of 'organic intellectuals' in Acehnese civil and political society, the wider effects of such projects on the production of knowledge may remain far-reaching for years to come. Inevitably, the international demand for local participation in such projects, typically in a designated 'junior' role, has centered on Banda Aceh, and served to reorient the focus of many a (full-time) lecturer, teacher, student, and organic intellectual in the aftermath of the tsunami. Their ranks were already depleted as the tsunami claimed large numbers of victims among campus colleagues and classmates, NGO activists and workers. Unsurprisingly, the post-tsunami deluge of international experts, advisors, consultants and technicians also allowed for large multi-donor budgets to trickle down to local academic and other researchers, thus contributing to the channeling of such energies in Aceh.

---

[22] Note, for example, World Bank/DFS, *Aceh Conflict Monitoring Updates* and related papers produced by the World Bank (Indonesia) available at www.conflictanddevelopment.org, and the numerous reports focused on Aceh in 'crisis' by the International Crisis Group available at www.crisisgroup/home/org.

Entangled Encounters

While the drive for a hypercentralization of relief and reconstruction efforts in early 2005 had powerful and perhaps lasting effects,
it never formed a seamless web for representatives of the Indonesian
government, international humanitarian organizations, or, indeed, affected local populations. Other competing dynamics also influenced
the parameters of interventions in early post-tsunami Aceh. First of all,
Indonesian political processes and institutions —ranging from the riva-
lry between the president and his vice president to the expanded legislative powers of the parliament and the enduring influence of the
military (including in business corporations) — exerted discernible influence on the nature and direction of responses to the disaster. Moreover, while mindful of their respective official mandates, and the
limitations thereof, international humanitarian organizations at times
demonstrated a capacity for constructive engagement with (local) official authorities, and, as a result, perhaps gained a surprising measure of
discretion in implementing relief and reconstruction assistance. Finally,
the relative weakness of organized civil society in pre-tsunami Aceh
notwithstanding,[23] there was early evidence of considerable activism
among survivors, who found themselves in a peculiar place where they
had, on the one hand, lost nearly everyone and everything, and thus
perhaps felt that they had little left to lose, while, on the other hand,
they had gained renewed recognition as "internally displaced persons"
with specific rights and needs under international humanitarian and
human rights law. In this regard, the return to Aceh of some of the best
and the brightest, who had left during the brutal repression of student,
human-rights, and other civil society movements in 2000–01, further
added to the "social capital" available for local affected populations in

---

[23] See Aguswandi, "Breaking the Deadlock: Civil Society Engagement for Conflict
Resolution," in Eva-Lotta E. Hedman (ed.) *Aceh under Martial Law: Conflict Violence
and Displacement*, RSC Working Paper no. 24 (Oxford: University of Oxford, 2005),
pp. 45–52.

post-tsunami Aceh. In short, it was not only government, but also politics, that contributed to shaping relief and reconstruction in post-tsunami Aceh.

As the novel modes of governmentality introduced ostensibly to improve the lives of displacees gained momentum, the wider and more dramatic transformation of a province re-emerging in the wake of protracted militarized conflict and sudden natural disaster also held out the promise of a new kind of politics, including challenges to dominant post-tsunami problematizations and representations of the 'internally displaced person.' Such challenges surfaced as Aceh once again leaned into the future of contested elections, anticipating the reordering of local political dynamics, including in many communities affected by displacement. They differ from the myriad project evaluations of 'successes, failures, and lessons learned' by pointing beyond the common terms of reference – or what may be referred to as the constitutive exclusions – inscribing projects and evaluations alike. At the same time, they direct attention to a meta-discourse on rights and responsibilities in situations of internal displacement, thus articulating a critical position 'in the name of the internally displaced.'

In the cases examined here, IDP collective action emerged in the context of this wider political shift in the province, articulating ways of belonging in the new Aceh. While framed in a discourse of rights familiar from the *Guiding Principles*, such challenges also pointed beyond displacement as a defining condition, reconnecting it to lived histories and transcendent futures through collective action. On the one hand, the right-to-return campaign recalled a murkier past in the present crisis, pointing to the continued legacies of local militarized conflict and violence, and reasserting old solidarities associated with the lived experience thereof in pre-tsunami Aceh. On the other hand, the barracks-protest campaign summoned a brighter future in the present emergency, stressing the novel opportunities for greater social justice, and affirming new socialities forged in the context of displacement in post-tsunami Aceh.

Conflict IDPs and the Right to Return

This promise of a new kind of politics was exemplified by the emergence of a campaign that challenged the very discourse of displacement in the aftermath of the tsunami in Aceh. As noted above, the circumvention of the IDP discourse in early post-tsunami Aceh pointed towards certain (and away from other) understandings of the nature of the 'problems' of displacement and the displaced. Indeed, the disregard for so-called 'conflict IDPs' and, moreover, for the complexity of—recurring, multiple, and/or secondary—displacement remained pervasive in international humanitarian practice into the post-Memorandum of Understanding (MoU) period in Aceh, and also served to reflect and reproduce official government discourse. In the context of wider political changes, however, thousands of Acehnese whose displacement pre-dated the tsunami launched a 'right-to-return' campaign, thus, in effect, articulating a critical challenge to the constitutive exclusion of 'conflict IDPs' from dominant humanitarian discourse and practice in the aftermath of the tsunami.

After several unsuccessful attempts by individuals and families to return to their communities in central Aceh, a collective effort was launched on 10 December 2005 by some 4,500 to 5,000 people who had previously found sanctuary from conflict and violence in Pidie and Bireuen, two adjacent northern districts.[24] As transportation promised by local government officials failed to materialize on the day it was scheduled, this planned collective return was transformed into a protest march. Taking their first stop at Abu Beureueh mosque in Beureuneun (Pidie), a prominent site of resistance and refuge alike in the social imaginary of many Acehnese, the marchers continued some ten kilometers on to Lumputot (Bireuen). There, they eventually boarded trucks and buses headed for the interior highlands of central Aceh, accompanied by yet more would-be returnees joining from other host commun-

---

[24] According to the Center for Humanitarian and Social Development (CHSD), a local NGO in Pidie working with the IDPs, some five thousand names were collected at the outset of the return march. Author's interview, Pidie, January 11, 2006.

ities in Bireuen, and stopping to set up large makeshift tent camps by the mosques at two locations known as Km 60 and Ronga-Ronga.[25]

This mobilisation of a collective return to the highlands by thousands whose displacement due to conflict and violence pre-dated the tsunami presented a critical challenge against the silence on "conflict displacement" inscribing international humanitarian practice and local political ambition alike. On the one hand, this protest action seemed to have caught the "international humanitarian complex" camped out in Banda Aceh in the aftermath of the tsunami by surprise, prompting a brief flurry of fact-finding missions and roundtable discussions, and, not least, considerable consternation about this departure from the more familiar script of the displaced as "speechless emissaries" of suffering, or grateful beneficiaries of humanitarianism.[26] As speculation of a less public nature focused on the hidden hand and political motivations *"behind"* this protest campaign, it is perhaps unsurprising that contemporary reports emphasized the humanitarian nature of the difficulties encountered by these would-be returnees, spotlighting the plight of people who had reportedly collapsed from exhaustion and starvation within the first week.[27]

On the other hand, there was ample evidence of other kinds of impediments to the safe return of these (former) conflict displacees, many of whom had fled their villages in the central highlands in 2000–2001 during a period of especially "high-intensity" conflict involving not only GAM (*Gerakan Aceh Merdeka,* Free Aceh Movement) and the Indonesian military, but also Javanese (trans)migrants and so-called an-

[25] See, for example, "Pengungsi Kembali ke Koloni" *AcehKita* (12-18 December 2005), p. 10 available at www.acehkita.com.
[26] Liisa Malkki, "Speechless Emissaries: Refugees, Humanitarianism, and Dehistoricization," *Cultural Anthropology* 11, no. 3 (1996): 385. For an early and critical formulation, see especially Barbara Harrell-Bond, *Imposing Aid: Emergency Assistance to Refugees* (Oxford: Oxford University Press, 1986).
[27] World Bank—DFS, "Aceh Conflict Monitoring Update, December 1–31, 2005," available at www.conflictanddevelopment.org.

ti-separatist militia groups.[28] Indeed, many of the difficulties encountered by would-be returnees suggested a ready local reaction and swift (extra-) political response to this collective campaign. First of all, would-be returnees were required to produce formal verification of "IDP status" with the signatures of local government officials in the village where they had taken refuge and in their home village, without any commensurate government guarantees of security on their return, thus catching these (former) conflict displacees in something of a double-bind. Moreover, local civilian government officials also intervened in ways that seemingly aimed at silencing any would-be collective "IDP voice" (for instance, by holding meetings without IDP leaders or other concerned parties present) and at undermining any united IDP front (e.g., by calling for further dispersals of IDPs to different sub-district locations and individual villages).[29] In addition, there were reports of intimidation and assault at the hands of the Indonesian military and police. On 8 January 2006, military and police entered the makeshift camp at Ronga-Ronga, and IDPs were made to board buses and trucks as uniformed officials shouted out destinations: Timang Gajah, Rimba Raya, Pinto Rime Gayo, all in Bener Meriah. Finally, there were reports of violence targeting returning IDPs and/or their property, as well as cases of fighting with local youths or former militia, in places where local leaders refused to provide security guarantees. In a village in Ketol (Aceh Tengah), for example, eight IDPs trying to secure the signature of the village head were beaten up and had their motorbikes destroyed by local residents.

---

[28] See, for example, Ali Aulia Ramly, "Modes of Displacement during Martial Law," in *Aceh under Martial Law: Conflict, Violence, and Displacement*, ed. Eva-Lotta E. Hedman, Working Paper no. 24 (Oxford: University of Oxford, Refugee Studies Centre, 2005), especially p. 18.

[29] Author's interviews with representatives from the Center for Humanitarian and Social Development (CHSD), the Aceh Monitoring Mission (AMM), the local government in Takengon, and with numerous conflict IDPs from Pidie and Bireuen at locations in Aceh Tengah, including at 60K, Ronga-Ronga, and Laut Tawar, as well as with other concerned observers in Banda Aceh, 10-15 January 2006.

These entangled encounters and the kinds of "friction"[30] they produced served to disrupt the dominant discourse on "IDPs" in post-tsunami Aceh by directing attention to the issue of (former) conflict and displacement. That is, having been displaced by conflict during the previous militarized campaigns of the counterinsurgency, IDPs who sought refuge with host communities had remained largely invisible in the wider context of early post-tsunami Aceh. While some 1,800 conflict IDPs were reportedly still in camps only days before the tsunami struck,[31] there was no corresponding figure for those who had sought refuge from militarization and violence with relatives or friends. This is not surprising, given the informality and fluidity of such arrangements compared to the official registration and regulation of IDPs in designated shelter areas. Nonetheless, conflict IDPs who had found shelter in host communities outnumbered by a considerable margin their encamped counterparts at that time.

In as much as the attempted collective return of (former) conflict IDPs targeted a particular mode of governmentality and displacement in post-tsunami Aceh, it also revealed something of the conditions under which such a critical challenge may emerge. In this regard, it is important to recognize the extent to which local political conditions shaped the horizon of "conflict IDPs" living in host communities during the early implementation of the MoU in Aceh. Not only did such conditions influence IDP efforts and prospects of recognition, but also their possibilities for return. For example, it was only in the aftermath of the peace agreement and its implementation — involving, notably, the demilitarization of host-community areas — that conflict IDPs in Bireuen and Pidie were able to organize their collective return to the central highlands of Aceh, a region formerly governed under a single administrative district, Aceh Tengah, which in 2003 had

---

[30] Anna Lowenhaupt Tsing, *Friction: An Ethnography of Global Connection* (Princeton University Press, 2005).

[31] International Organisation for Migration, "Update on the IDP situation in Aceh," 20 December 2004. See updates available at www.iom.or.id/news.

been divided by the (contested) creation of a second district, Bener Meriah. Moreover, since the Cessation of Hostilities Agreement (COHA)
had broken down in 2002 following an attack on the Takengon local
office of the international monitoring body, it was perhaps no coinci-
dence that these conflict IDPs had remained displaced for so long. As
noted above, their original flight from violent conflict had taken place
as long ago as May or June 2001. Nor was it surprising that IDPs eventually seized on collective action in their efforts to secure a return to
central Aceh, where so-called militia groups had gained particular noto-
riety. These groups were significantly not included in the MoU, but
they still enjoyed the backing of local businessmen, as well as civilian
and military officials.[32]

If the demilitarization of host-community areas allowed for
new forms of collective action by (former) conflict IDPs in Pidie and
Bireuen, a political campaign advocating the division of Aceh unders-
cored the urgency of returning to the central highlands lest administrative remapping were to render such return even more uncertain.
Indeed, the timing of the attempted collective return movement on
December 10 followed on the heels of a 4 December 2005, rally in Jakarta, when seven local *bupati* (regent) unilaterally declared their seces-
sion from Aceh and their proposal to establish two new provinces:
"Aceh Leuser Antara" (ALA), which would be forged out of the districts
of Aceh Tengah, Aceh Tenggara, Aceh Singkil, Gayo Lues, and Bener
Meriah, and "Aceh Barat Selatan" (ABAS), which would be formed
from Aceh Barat, Aceh Barat Daya, Aceh Jaya, Nagan Raya, and Si-
meulue.[33] While less immediate, the scheduled termination of the EU-

---

[32] See, for example, Imparsial Team, "Report Post-MoU Monitoring on Aceh, Aug.
15–Oct. 15, 2005," October 2005. With the establishment of the Aceh Peace-
Reintegration Agency (BRA) in February 2006, anti-separatist groups were included
as beneficiaries eligible for government-funded reintegration programs, as were "conflict-affected persons" in general, and also GAM supporters who had surrendered
prior to the MoU.

[33] "Pemekaran Aceh Idak Jadi Prioritas", 7 December 2005, available at
www.acehkita.com.

led Aceh Monitoring Mission in mid-March 2006 also threatened the prospects for safe return in what remained, in many respects, an unreformed outpost of the "new" Aceh.[34]

Such were the conditions facing displacees from earlier rounds of (para-) militarized conflict in the Central Highlands as they mobilized a collective return movement in December 2005. In doing so, it is important to recognize, they inscribed their protest within a distinctly universalist language of legitimation resonating with the *Guiding Principles*. In this vein, they also posed a critical challenge against the efforts at containment of the very universalist aspirations held out in the *Guiding Principles* in their application to tsunami but not conflict-related displacement in Aceh.

Tsunami IDPs and Barracks Relocation

A second example of this promise of a new kind of politics emerged with another IDP campaign targeting the governmentality of displacement. The promotion of so-called "temporary relocation shelters" or "barracks" in early post-tsunami Aceh also left enduring legacies on the nature and direction of efforts to improve and regulate the lives of displaced populations. As the "international humanitarian community" was drawn into the orbit of various "solutions" (but not others), moreover, those solutions gained added momentum. In this regard, the materialization of barracks as a preferred mode of governmentality anticipated the institutionalization of needs assessments and aid delivery focused on such camp-like relocation complexes. However, the introduction of novel government regulations on resettlement in 2006 served to differentiate among barracks populations according to (unequal) pre-tsunami property relations. As the news of such regulations spread precisely at a time when the much-awaited local election campaign held out the promise of (equal) political franchise, it is perhaps

---

[34] In mid-March, the AMM cut back its staff of 220 monitors by almost two-thirds, but also agreed on a three-month extension with the government. After a further extension, the AMM finally departed from Aceh in December 2006.

unsurprising that the barracks themselves emerged as a new field of struggle for the displaced.

With tens of thousands displaced by the tsunami still residing in such "temporary location shelters" in 2006, the mobilization of protests centered on the barracks faced quite different circumstances from the collective campaign to return to the Central Highlands discussed above. On 11 September, an estimated two thousand people joined in a major demonstration at the offices of the BRR (Badan Rehabilitasi dan Rekonstruksi [NAD-Nias], or the Rehabilitation and Reconstruction Board) in Banda Aceh to refocus attention on the situation facing the IDP barracks population. According to reports, this first demonstration remained entirely orderly and resulted in a negotiated agreement to review BRR regulations affecting the resettlement of what, in fact, has remained a considerable proportion of the barracks population— renters, squatters, and the landless.[35]

On 19 September, and continuing until the following day, the BRR once more became the target of a major demonstration focused on government regulations and the resettlement of barrack IDPs. This time, the demonstrations, which again involved an estimated two thousand people, also featured a blockade of the BRR. As the police moved in to disperse the crowds, some protesters reportedly threw rocks in their direction, damaging a police car.[36] According to the BRR, three attempts at negotiating a settlement with the leaders of the second demonstrations failed, as demands had escalated "to include unrealistic requests," including the transfer of money into an NGO account. However, aside from "some loud, well-orchestrated cheering and chant-

---

[35] See, for example, BRR International Update, "Demonstrations at and Blockade of BRR in Banda Aceh," 20 September 2006, available at www.e-aceh-nias.org.

[36] See, for example, "Protesters Attack Aceh Tsunami Reconstruction Office," *Reuters*, 20 September 2006. According to Banda Aceh deputy police chief Dede Setyo, the police "decided to disperse the crowd because they had been staying outside the BRR offices beyond the timeline that we gave them."

ing," the BRR concluded in its update on these events, "both demonstrations have been conducted peacefully."[37]

As noted above, these demonstrations emerged against the backdrop of new regulations on resettlement issued by the BRR in June 2006. In brief, these regulations made new provisions for displaced populations not covered by the resettlement schemes developed for those who had owned their land before the tsunami, that is, for renters, squatters, and the landless. According to contemporary estimates, these groups comprised the majority of the 70,000 to 100,000 people displaced by the tsunami who were still living in some 150 government barracks located across different parts of Aceh.[38]

As the new government regulations of June 2006 served to "unpack" and differentiate between the displaced on grounds of pre-tsunami property relations, they also, invariably, set into motion new forms of contestation over the nature and direction of post-tsunami resettlement. In important respects, they thus reflected and reproduced a shift away from the "humanitarian needs" discourse and practice within which the barracks — and their residents — had remained inscribed since the early post-tsunami months of 2005. That is, as the new regulations made distinctions among different categories of displaced persons, many of whom had encountered each other as "IDPs" in the context of the barracks, it was not merely "confusion" and a "simmering crisis" that resulted, but also the opening up of a space for a new politics of recognition.[39]

This new politics of recognition gained momentum as a result of the local elections on the horizon and the voter registration drive

---

[37] BRR International Update, "Demonstrations at and Blockade of BRR in Banda Aceh," available at www.e-aceh-nias.org.

[38] In September 2006, BRR cited the figure 70,000 (in BRR," Special Unit on Barracks") and 100,000 (in "BRR International Update"). See BRR International Update, "Special Unit on Barracks," September 2006.

[39] The "confusion" and "simmering crisis" are noted in Oxfam International, "The Tsunami Two Years On: Land Rights in Aceh," *Oxfam Briefing Note*, (Oxford: Oxfam, 30 November 2006), p. 7.

underway in 2006, including among IDPs in barracks and elsewhere. In the context of a highly contested election campaign, with local parties and so-called "independent candidates" running against the national party machines of Golkar and others, widespread collective action re-emerged, with local communities and groups further expanding an already impressive repertoire of protest. Such protests have focused on a range of reconstruction-related issues, and have targeted not only the BRR, as noted above, but also the BRA (Aceh Peace-Reintegration Agency), which was established in February 2006, with government funding to establish reintegration programs whose beneficiaries would include (former) militia groups, GAM supporters who had surrendered prior to the MoU, and "conflict-affected persons" throughout all rural communities.[40] Protests have also directed attention to the electoral process and, in places, the outcome of the elections.

While it is hardly surprising that the BRR and the BRA — the key coordinating and implementing government agencies focused on post-tsunami and post-conflict reconstruction — have been targeted by IDP protests, it is noteworthy that, in the post-election months of 2007, the local parliament in Banda Aceh emerged as a new site of collective action in the name of the internally displaced. Within days of another round of protests involving hundreds of IDPs and, once again, Forak (the Inter-Barrack Communications Forum) at the BRR on 9 April 2007, the Acehnese Alliance of Youth and Students (Alee) called for an international audit of the BRR in a demonstration at the parliament in Banda Aceh. A few months later, on 30 July, the parliament building was occupied for some thirteen hours by (former) conflict IDPs from Bener Meriah and Aceh Tengah, where tensions over the

---

[40] Although the politics of reconstruction and aid delivery demands more careful attention than is possible here, the assessments produced by donors and "partners" frequently mention the following "problems": the slow pace of delivery, the lack of transparency in identifying recipients, the absence of checks on contractors, and the corruption of funds.

mismanagement of reintegration benefits by the BRA have remained acute, and where the elections were marred by comparatively high levels of fraud and violence. In this vein, challenges in the name of the internally displaced have joined the wider repertoire of protest in the new Aceh.

New Beginnings

As the broader political landscape underwent dramatic transformation in the aftermath of the December 2004 tsunami, some of the earlier concerns for the protection and the rights of internally displaced persons perhaps seemed to have been unwarranted in the first place, or, at the very least, now appeared dated to some observers and practitioners. However, as suggested above, the containment of the IDP discourse and the promotion of temporary relocation shelters in early post-tsunami Aceh pointed towards certain (and away from other) understandings of the very nature of the "problems" of displacement and the displaced. While not without critics from the outset, these new modes of governmentality focused on displacement met their most critical challenges through the entangled encounters with the very (excess) populations who, in important respects, rejected their own constitutive exclusion (as 'conflict IDPs,' or as 'squatters, renters' etc.) from the schemes adopted for the ostensible improvement and regulation of the displaced.

These encounters also revealed, in word and deed, a deeper critique of such schemes as cogs in the wheels of the "anti-politics machine" of post-tsunami reconstruction. That is, as actually existing displaced persons evoked the *Guiding Principles* as a "public transcript" of sorts, identifying the proper roles and relations of concerned parties in situations of internal displacement, they also challenged national and local government officials, as well as international humanitarian organizations, to recognize the promise held out by a universalist (trans)national discourse of rights and responsibilities. Having presented a critical challenge to the intelligible field of post-tsunami displacement problems, these protests also anticipated the search for new solutions, as evidenced by the search for reforms of existing schemes to improve and regulated the lives of the displaced in post-tsunami Aceh.

On the one hand, the mobilisation of a collective campaign demanding the "right to return" for (former) conflict IDPs evoked an alternative narrative of pre-tsunami Aceh, recalling a politics of displacement and belonging fueled in part by militarized conflict and violence. Indeed, in challenging the governmentality of displacement in the language of (de)legitimation offered by the *Guiding Principles*, the thousands formerly displaced by conflict in the Central Highlands also, in effect, ruptured the (empty) categories of 'IDP' and 'post-tsunami' through their very insistence on the past in the present crisis. On the other hand, the emergence of the barracks as a site of contestation presented another kind of difference against the governmentality of displacement in Aceh. That is, in challenging the shift towards a policy of differentiation among barracks populations, once again in terms resonating with the *Guiding Principles*, protests centered on the barracks and their populations, in effect, focused attention on larger questions of social justice and property relations. While aimed at government regulations for resettlement in the post-tsunami period, these protests also suggested a revisionist critique of a more far-reaching political kind, rejecting a return to the ex-post-ante of pre-tsunami social and economic relations on principle. In distinct ways, then, the protests examined here served to rupture the expert discourses that have inscribed humanitarian relief and reconstruction efforts in the aftermath of the tsunami that hit Aceh on 26 December 2006. In this vein, these protests also point beyond the production of knowledge anchored in a governmentality of displacement.

As suggested above, a focus on protest can prove illuminating in exploring the conditions under which governmentalities of displacement encounter critical challenges that they cannot fully contain. Moreover, it directs attention to moments when the targets of so-called "expert schemes" articulate their own critical analysis of the situations with which they are faced. This line of inquiry about the dynamics of displacement and protest thus recasts familiar questions of 'agency,' 'voice' and 'participation' in the field of refugee studies. Finally, an examination of particular forms of contestation, such as protest, also helps identify the kind of friction produced when something new is emerging, that is, when "a difference is introduced into history in the form of politics."

Eva-Lotta E. Hedman is Senior Research Fellow in the Refugee Studies Centre, Department of International Development, University of Oxford. Her recent related publications include *Conflict, Violence and Displacement in Indonesia* (ed.) (Cornell University: Southeast Asia Program Publications, 2008).

The Politics of Reconstruction, Gender, and Re-Integration in Post-Tsunami Aceh[1]

Jacqueline Aquino Siapno

Abstract

This paper examines the agency and political subjectivities of marginalized groups, especially rural women, in post-tsunami policies and processes of Gerakan Aceh Merdeka (GAM) re-integration into civil society, reconstruction, rehabilitation and peace-building. It argues that Demobilization, Disarmament and Re-integration programs and Security Sector Reform which does not include the participation of female ex-combatants, women ex-veteran (broadly defined) supporters of GAM, and stronger and more democratic civilian oversight mechanisms -- is a policy that is set-up to fail, and may result in longer-term security problems. The paper also examines the paradox of increasingly numerous International Non Governmental Organisations (who are predominantly secular and tend to avoid 'Islam') working on "gender" in Aceh and the increasingly oppressive mechanisms set-up by religious institutions and leaders to constrain women's voices, participation, and movements in Aceh.

[1] I owe a debt of gratitude to Fatimah Syam, Director of LBH APIK, Lhokseumawe, Khairani SH (RPUK), Darwis, SH, Suraiya Kamaruzzaman, Muhammad Nur Djuli, Teuku Kamaruzzaman, Saiful Mahdi (Aceh Institute), Ahmady (Aceh Kita), Kym Holthouse (formerly of UNDP Aceh), and Lilian Fan (Oxfam, Aceh), for making my research trips to Aceh both productive and enjoyable. I wish to thank Malathi de Alwis and Eva-Lotta Hedman for inviting me to be part of this project and for commenting extensively on drafts of this paper.

Writing about "reconstruction" in war, conflict and so-called "post-conflict" situations, Deniz Kandiyoti, identifies the following multiple transitions in the process of reconstruction: a security transition -- from armed conflict to sustainable peace, from non-statutory forces to the establishment of "modern" police and military institutions; a political transition -- from leadership of ex-guerrillas to the formation of a nation-state; and a socio-economic transition -from a "conflict" economy to more "sustainable development."[2] While Kandiyoti focused on Afghanistan, this analysis is very much relevant to the situation in Aceh today, especially in relation to a comparative analysis of the paradoxical rise of both Islamism and "gender reconstruction programs" in both countries.

The narrative of hard-line, male-dominated militarism and predominantly male nationalist armed struggles betraying women's struggles for political justice is nothing new – in fact it is continually repeating itself, produced and re-produced in many parts of the world, such as in Afghanistan, Palestine, Sri Lanka, the Philippines, East Timor, and Nepal."[3] In Aceh, however, the contradictions between women's agency and militarized masculinities amount to an everyday "civil war" – played out in the home, the family, on the streets, in criticisms of lack of inclusiveness of the peace process, in the Mahkamah Syariah courts, in protests over greater accountability and egalitarian processes in reconstruction and rehabilitation programs.[4] In part, this "civil war" recalls problems of post-conflict transitions elsewhere: "Ironically, revo-

---

[2] Deniz Kandiyoti, "The Politics of Gender and Reconstruction in Afghanistan," UNRISD Report, 2006.

[3] See for example, UN Police Dan Karki, "Cosmopolitan Militaries: Challenges in Training Police in Gender and Human Rights – Comparative Perspectives from Nepal and Timor Leste," Paper presented at a Workshop on "Security Sector Reform: Gendered Perspectives," Delta Nova, Dili, Timor Leste, 3 May 2007.

[4] See for example, Raihana Diani, "Women's Roles after the Tsunami and Conflict: Challenges and Hope," Paper presented at a Conference on "The Future of Aceh: The Remainders of Violence and the Peace Process in Nanggroe Aceh Darussalam," Harvard University, 24-27 October 2007.

lutionary interpersonal relations are sometimes mimetic of state repression, operating both in response to, and as a reflection of, the logic of the same political repression that the revolutionaries were suffering at the hands of the government which they were trying to overthrow."[5] In this paper, I wish to focus on the individual agency and political subjectivities of marginalized people, especially women, who are often left out of re-integration, reconstruction and rehabilitation processes in post-tsunami Aceh.

Left-out of Re-integration, Reconstruction and Peace-building

In my interview with a prominent Acehnese humanitarian aid worker and scholar, one of only two women (from the Acehnese side) who served on the Rancangan Undang Undang Pemerintahan Aceh (RUU PA) committee on the law governing Aceh, Khairani SH argued:

> *Inong balee?*[6] GAM themselves do not recognize and value the contribution of these women in the armed struggle. They are not included in the peace-building processes, nor reconstruction, and rehabilitation. They do not have any power to be involved in decision-making processes. They are like low level corporals – who are told to go there, come here – but have never been allowed to choose and determine for themselves or for GAM. In all of these processes – from peace-building, transi-

---

[5] See Nancy Scheper-Hughes and Phillipe Bourgois (eds.), *Violence in War and Peace: An Anthology* (Malden: Blackwell Publishing, 2004), pp.18-19.; also Pierre Bourdieu, "Gender and Symbolic Violence," in *Masculine Domination* (Stanford: Stanford University Press, 2001); and Pierre Bourdieu and Loic Wacquant, "Symbolic Violence," in *An Invitation to Reflective Sociology* (Chicago: the University of Chicago Press, 1992).

[6] Inong balee refers to both widows (who lost their husbands during the armed conflict) and female armed combatants in GAM. The number of Inong balee varies, depending on which group (GAM or non-GAM) one speaks to, sometimes encompassing also women supporters, couriers, women hiding and protecting GAM in the villages.

tion, reconstruction and rehabilitation, institution building –
none of them have been included. Maybe one token, but not
enough. Even in other processes – when GAM made suggestions for the RUU PA – are there any women recommended by
GAM? Ngak ada![7]

These women are also victims – and should also be entitled to
justice, truth, economic recovery, trauma counseling – but always the priority is on GAM-the men. The women are not included at all. Those who have been receiving monthly
compensation, land, and other benefits – most of them are men.
A few women who were previously arrested, imprisoned have
been assisted. But the larger proportion of women (and we are
talking about a lot) have not been included. During Darurat
Militer they could not have access to humanitarian aid because
they are classified as GAM (even though it is their husband
who is GAM, the wife is not a member, but suffers the consequences.) Now during "peace-time", they get nothing. Apparently because they are just "wives", not GAM members. But it is
these women who felt the severe consequences of this armed
conflict. Why aren't their names included in the lists of people
who deserve support and entitlement? In the past, they were in-
timidated and stigmatized. And now, their contributions are not
recognized, nor are they allowed to participate. This is going to
be a problem in the future.[8]

Much has been written about the conflict and violence between
Gerakan Aceh Merdeka (GAM) and the Tentara Nasional Indonesia
(TNI), but the war within GAM remains under-examined. Several
highly respected, influential, and intellectually subtle women activists,
scholars, and lawyers in Aceh commented [in 2006]: "If GAM is in
power one day and rules this land, which is now highly likely as they
are about to form a political party…we'd like to create another country,

---

[7] Interview with Khairani, SH, Secretary-General of Relawan Perempuan Untuk Kemanusiaan (RPUK) and member of Rancangan Undang Undang Pemerintahan Aceh
(RUU PA), Aceh, March 2006. She is one of only two women members formulating
the RUU PA governing law for Aceh.
[8] Interview with Khairani, SH.

or at least move to another country."[9] When asked why she/they thought that, she replied: "We do not want to be ruled by "*orang-orang sakit*" (wounded people; but it can also pejoratively mean "*gila*" or insane), who have only known pain, torture, and war, and whose ways of solving problems is primarily through militarized, masculinist, violent means – without consulting with women."

I interviewed several GAM people asking them if there was a space to discuss critical comments such as the woman's perspective above, and while some of them responded "of course, there is space," others categorically said: "No – not at the moment – because we don't really have any power to do much. So it is not fair for Acehnese women and scholars to criticize us now." What is women's relationship with GAM? And what is their critique (or inability to critique in some contexts) militarism and hyper-masculinity (of both GAM and TNI) The answers to these questions include: The suppression and subordination of female voices by their own partners and spouses who are GAM leaders (this includes the prominent example of a female politician of Inong Balee who is also now the spouse of the Head of Badan Reintegrasi Aceh [BRA]); the generation gap between the GAM "old boys network" and young women leaders who often courageously criticize the "wise old men", but are then "lectured upon" in a patronizing way as "women who do not know their place";[10] and powerful GAM leaders "co-opting" outspoken Acehnese women, for example, "this problem is still within the family, between a brother and sister, an "uncle" and a "niece"...why did you have to expose it externally?." (There are several examples of turning public/political relations into internal/family relations, but due to the continuing fragility of peace, I will not mention names here.)

---

[9] Confidential interview with humanitarian aid worker and human rights activist, Aceh, March 2006.

[10] I have witnessed this happen a few times.

International organizations presumed to provide better models for democratization, participatory processes, and gender equality have not been exemplary. For example, the Aceh Monitoring Mission (AMM) has also been criticized for not being more inclusive of women's ideas in the peace-building and reconstruction processes in Aceh:

> AMM invited us to discuss gender and women's issues – but only now, when they are just about to leave. Before, when they had more time, they never invited us. Peace-keeping? Now that they are leaving, then they invite us. Most of them have left already. Why did they suddenly become so interested in gender and women before departure? Most likely because they have to fulfill some bureaucratic procedures. They asked us to write a report quickly and as soon as possible. I told them: fine, but don't you have to go home now?[11]

My research in post-tsunami Aceh began in March 2006, and I have worked closely with Ms. Fatimah Syam, Director of LBH-APIK, to conduct interviews in Lhokseumawe and Banda Aceh. This paper is thus primarily based on oral history interviews conducted during my research trips to Aceh from March 2006 onwards, from symptomatic and rhetorical readings of International Non Governmental Organisation (INGO) and Local Non-Governmental Organisation (NGO) reports, human rights reports, women's magazines, and from participant-observation in peace-building processes in Aceh.

No Political Justice or Legal Representation

A large number of new research reports have been produced on people's access to justice and other "reconstruction needs". A more symptomatic reading of the research reports, however, indicates a common pattern and uniformity of language – the language of development and reform — being used in all.

For example, the current discussion/s on a "truth and reconciliation commission" reveals this uniformity of language – whether it is

---

[11] Interview with Khairani, SH, *op. cit.*.

the rhetoric of the professionalized human rights NGOs or that of GAM. It is within the broader context of demands for "political justice" as a condition for sustainable peace in Aceh that I situate my own analysis of women's political subjectivities, female agency, and gendered perspectives on militarization and NGO and INGO reconstruction post-tsunami.

What do we do with subjectivities that do not speak from a place of trauma but speak beyond trauma? Engaging with Gayatri Spivak's question in her article: "Can the Subaltern Speak?", Sylvia Tiwon argues that the subaltern does speak, but we are not equipped, we do not have the capacity, patience, nor training to listen. She argues that subaltern women, in particular, do speak – but no one listens.[12]

So many of the lived experiences of violence and displacement remain hidden, stories remain suppressed, and their authors/actors/sources dis-appeared (not in the material, physical sense, but in an epistemological sense – either through self-censorship or "symbolic violence.") Bourdieu writes about "Symbolic Violence" – the forms of obligation and compulsion that are never recognized as coercive but experienced as generosity, as "aid", solidarity, as piety, personal loyalty, or self-respect. It is an invisible kind of coercion: the essence of which is that it is never recognized as violence as such, but is experienced as a legitimate system of governance and morality.[13] Tiwon observes that this silencing effect is akin to the subaltern's unfulfilled speech act. These articulations – "speaking with a great deal of trepidation" and in "whispered confidences", Tiwon argues, gives us a very dif-

---

[12] Sylvia Tiwon, "Narratives of Women's Experiences," Paper presented at a Conference on "Aceh Peace Process," Harvard, October 2007. See also, Gayatri Spivak, "Can the Subaltern Speak?," in *Marxism and the Interpretation of Culture*, ed., Cary Nelson and Lawrence Grossberg (Urbana, Il: University of Illinois Press, 1988), pp. 271-313. Timothy Mitchell makes a similar point in his chapter, "Nobody Listens to a Poor Man," in *Rule of Experts: Egypt, Techno-Politics, Modernity* (Berkeley, Los Angeles and London: University of California Press, 2002), pp. 153-178.

[13] Bourdieu, *op. cit.*

ferent idea of the shape of violence. "Women do speak ... but do we have the capacity to listen? Or do we hear only that which we wish to hear? Women's narratives become 'un-intelligible noise' – in which the unintelligible disorder is described as 'order'."[14]

In the context of Aceh, the question of who speaks, of what and to whom is one which requires closer scrutiny. Is there really democratic space in the present moment to criticize GAM and their violations? Based on extensive interviews with women's groups and widows, international security advisers, and my own participant-observation of how elite GAM leaders react to criticism, my assessment is that there seems to be very little space to criticize GAM in the present juncture. Instead, there is a tendency, (common amongst former non-statutory guerrilla forces undergoing political transition to leadership in state apparatuses) towards authoritarian mentality and attitudes.[15] By contrast, Acehnese women's rights workers interviewed by the author have been much more profoundly critical of the representations and interpretations of the causes of violence and forced displacement in Aceh. In confidential interviews, they have been outspoken in their criticisms of GAM. These criticisms however, remain much more muted in public.

For example, in contrast to the Komnas Ham, Kontras, and LBH Banda Aceh reports on general human rights violations, several women's groups interviewed cited a third reason for political violence and displacement. That is, they attributed such phenomenon, at least in part, to GAM's political strategy of village evacuations.

> During the UN Special Rapporteur's investigation especially on women and IDPs, one of the women humanitarian workers here outspokenly commented in public: some of the mass displacement is not because of fear, but because they are forced to

---

[14] Tiwon, *op. cit.*

[15] For a brilliant un-packing of this kind of mentality, see for example, James Scott, "The Revolutionary Party: A Diagnosis and a Plan," in *Seeing Like a State: How Certain Schemes to Improve the Human Condition have Failed* (New Haven: Yale University Press, 1998).

do so, in the interests of GAM's political strategy – GAM told village people to evacuate. What happened to this woman after this statement? She could not leave the airport without being escorted by PBI (Peace Brigades International). GAM was on her trail. They put her organization on a blacklist.[16]

These lived experiences do not make it to the independent human rights reports, the nationalist GAM reports, or the international "conflict resolution" and "crisis" reports either because they do not fit the scripted, predictable, formulaic categories of human rights narrative rhetoric, are too politically-incorrect and undermine both the Indonesian nationalist and Acehnese nationalist paradigms, and/or are perceived to be "in-authentic subjects" in the politically-correct style of much conflict, violence, nationalist, and crisis writing. More importantly, silences, unwillingness to respond, and the sheer inability to narrate or express grief in questionnaires and interviews… also has its pleasures…it is also a long-term strategy, and skill, of care of the self as a practice of freedom. For many of the women interviewed for this paper, silence is a cultivated skill and a choice (not something necessarily negative and enforced), which can assure anonymity and continued mobility, ability to go in and out, create horizontal and vertical networks, and live in Aceh without having to be evacuated, exiled, displaced. In certain political contexts, such as the current one in Aceh, certain lived experiences are left at that – lived, but not named (written about, with identifiable sources, and reported publicly). Not all experiences and communities are accessible – and for some, words fail to describe, as the vocabularies we have for some of these experiences are limited.

There are several cases…reported by women colleagues…where it is very difficult for the women to advocate and seek access to justice for what they have experienced. They think of the effect

---

[16] Confidential interview, Chairperson, Women's Rights group in Aceh, March 2006.

not only on them, but on their families. Especially if they are seeking legal justice. People are killed, people are disappeared...victims and their families have to think very hard before they make a move – to publicly open up their case, to ask for advocacy support. If you look at it from a sense of ' resisting', this could take different forms: they suppress their voices, lessen their activities. A large proportion of them leave – to a safer place.[17]

Armed conflict and displacement changed gender relations and sexual politics in ways that would conventionally be unacceptable in times of peace, such as for example men using the war as an excuse not to go to work, and TNI soldiers using teenage women in the villages as *pemuas nafsu* (objects to satisfy their sexual desires).

Because of the armed conflict, peasant farmers cannot work on their fields. After they come back from displacement, their harvest is gone, ruined. Their livestock are dead. Sometimes their plots of land have been taken over by someone else. They have lost a lot not only economically, but socially and psychologically as well. Our way of life – our social practices of spending time mengaji di balee – all of this is gone. Even gender roles are reversed: men and husbands use the armed conflict with TNI as an excuse not to economically support their wives (menelantarkan), claiming it is the woman, the wives who have to look for a living as it is too dangerous for men to go outside.[18]

On "psycho-social self-reconstruction" and "emotional healing", there is no shortage of documentation and excellent research.[19] However, a more careful reading and analyses of the reports tend to presume that some form of external psychiatric or biomedical interven-

---

[17] Interview with Khairani, SH, *op. cit.*

[18] Interview with Director, LBH APIK – Lhokseumawe, March 2006.

[19] See, for example, Byron Good et al., *Psychosocial Needs Assessment of Communities Affected by the Conflict in the Districts of Pidie, Bireuen and Aceh Utara, 2006* (Harvard Medical School Department of Social Medicine, IOM Aceh, Universitas Syiah Kuala, Bakti Husada, September 2006). (See also Part II conducted in other districts).

tion is necessary in order for emotional healing to happen. What seriously needs to be researched, analyzed and discussed are local knowledges and forms of resilience that do not need expensive "rule of experts" and "external interventions". If we only tend to see problems in Aceh as a "lack of ..." this and that, we will never achieve any sustainability in the long-term. A majority of the reports, from the World Bank report on "GAM Integration Needs Assessments" to "Psychosocial Needs Assessments" are coming from the perspective of "lack" or "complete absence of" any existing local beliefs, practices and institutions that can provide the cure to societal ills in Aceh. These produces a circular problem creating a culture of dependency in which external/foreign experts identify "needs" to justify their continuing presence in providing reconstruction and reintegration support. Another problem is that, while there is a dearth of "psychosocial counseling" for tsunami-victims, "emotional healing" programs have yet to be introduced to armed conflict victims and their families, who to this day remain clouded in mystery as to the fate of their kidnapped, disappeared, and extrajudicially executed loved ones. For them, unlike the tsunami victims, recovery and tawwakal may not happen for a very long time, due to the as yet tenuous and volatile lack of acknowledgement of crimes against humanity committed against Acehnese and access to justice and truth and reconciliation commissions yet to be set-up. As Geoffrey Robinson writes, "The past cannot remain shrouded in mystery. In such situations the victims continue to seek justice and are unable to come to terms with their sorrow and sadness."[20]

---

[20] Geoffrey Robinson, "East Timor 1999: Crimes against Humanity," a Report commissioned by the United Nations Office of the High Commissioner for Human Rights (OHCHR) (Los Angeles: University of California, July 2003).

Vocabularies of "Power" in Post-Tsunami Reconstruction

Darini Rajasingham-Senanayake's article, "Sri Lanka and the Violence of Reconstruction"[21] also speaks to the current situation in Aceh. Some of us do experience the representation of reconstruction in Aceh – in glossy brochures, sexy visuals, cut-and-paste universalizing languages (but with some so-called "Aceh culture-specific" points to make it look "authentic") – as symbolic violence. This is particularly true when one examines the highly "organized", well-funded, hegemonic governmental systems of rich INGOs versus the fragmented, heterogeneous, "not-so-organized" people in the *gampongs* (villages).

As Edward Said claims about Orientalism: "It not only creates but also maintains: it is, rather than expresses, a certain will or intention to understand, in some cases to control, manipulate, even to incorporate, what is a manifestly different world."[22] The current scholarship on 'Islamic fundamentalism,' notes Roxanne Euben, is also an exercise in power:

> The power to construct and control a subject that has little opportunity to contest either the interpretation or the terms of the discourse; the power to dictate the parameters of the field, from which experts regularly pronounce the identity, meaning, and function of a movement without reference to the adherents' own understanding of the connection between action and meaning.[23]

"Re-construction", according to several people I interviewed in Aceh (January and February 2007) and last year (March 2006) has primarily been interpreted by donors and implementing agencies alike as "physical reconstruction" of infrastructure and housing. Alongside this physical reconstruction of houses, what we are also currently seeing and experiencing not only through abstract representation but embo-

---

[21] Darini Rajasingham-Senanayake, "Sri Lanka and the Violence of Reconstruction," *Development* 48, no. 3 (September 2005): 111-120.
[22] Quoted in Roxanne Euben, *Enemy in the Mirror: Islamic Fundamentalism and the Limits of Modern Rationalism* (Princeton, NJ: Princeton University Press, 1999).
[23] Euben, *op. cit.*

died, corporeal experience, is the implementation of what Muhammad Iqbal, the Pakistani political theorist, calls the "reconstruction of religious thought in Islam."[24] In his book, which bears the same title, Iqbal explores the historical meaning/s of *ijtihad* (which has also been interpreted as a kind of rhetorical form of *jihad.*) The debate on women in Islamic societies has been ongoing for some time. Several Muslim theologists and feminists are grappling with this issue at the level of the "pure philosophy" of the Quran and questioning the legitimacy of the Sharia and H*adith* by exercising *ijtihad* (critical interpretation). Iqbal defines *ijtihad* (from *jahada*, to exert) as the principle of movement in the structure of Islam: "In the terminology of Islamic Law it means to exert with a view to form an independent judgment on a legal question. The idea . . . has its origin in a well-known verse of the Quran – 'And to those who exert we show our path'."[25]

"Re-construction" has also been used by Acehnese women's rights advocates to refer to the "un-reconstructed men and leaders" who dominate politics and the public sphere in Aceh, and who, in spite of all the rhetoric on "The New Aceh", seem to have no plan or intention to re-construct themselves and their attitudes and behaviours towards women and "gender". And then, there is the use of "psychosocial re-construction" – ostensibly the re-construction of an individual or collective psyche in the context of the social – as part of recovery from trauma.[26] Taken together, the re-construction of Islamic thought (the so-called "new Islamist movements") and the now obligatory self-and-collective re-construction of women's behaviour, mentality, mobility, and space – is interpreted by critical observers as an explosive Molotov cocktail. If Aceh has been a strongly Islamic region over the longue duree, prior to colonization, how is this new Islamist movement different?

---

[24] Muhammad Iqbal, *The Reconstruction of Religious Thought in Islam* (Lahore: Institute of Islamic Culture, 1986).

[25] *Ibid.*, p. 117.

[26] See for example, Good et al., *op. cit.*

For one, it is alleged to be an initiative of the central Jakarta/Indonesian state bureaucracy, it is well-funded and well-resourced and continually expansive, and it is interpreted by some Acehnese as a "revenge" (with a twist) towards the "bad un-reformable, rebellious Acehnese child".

How do women negotiate such a multi-layered minefield? What is GAM's relationship with the new Islamist movement emerging post-tsunami? When asked why there seems to be mounting pressure on women to adopt outward expressions of Islamic public piety (especially veiling), an Acehnese friend replied, "The big difference is the tsunami. People are looking for answers as to why the tsunami happened. There was a lot of soul-searching. Somehow spirituality filled the void. Then radical Islam became the most organized and assertive in providing the answers: for example, 'the tsunami happened because of women's sins', and so on."[27] Several banners (*spanduk*) on the tsunami and women's sins and other banners on morality and immorality were posted throughout Banda Aceh. There are lots of DVDs, cassettes, booklets, and other media and new media distributed and sold in the markets in Aceh on the theme of the tsunami and morality (in particular women's morality). There is a production of rhetoric on "women's sins" (focusing in particular on bodily comportment, female sexuality, and mobility) within the space/realm of the Islamist interpretations and representations of Islamic values.

Somehow, after the tsunami, the former GAM's primarily nationalist-secular leadership seems to have been put in a reactive, rather

---

[27] Conversation with Aguswandi, Post-Conflict Recovery and Integration Adviser, IOM, January 2007, Banda Aceh. See also his other articles in *Jakarta Post*: "Say No to Conservative Islam," 30 August 2006 and the fascinating response by Zulkieflimansyah, "Islam, Muslims and Democracy in Indonesia," *Jakarta Post*, 12 September 2006 (Zulkieflimansyah is a Member of The House of Representatives from PKS). See also Aguswandi's articles: "Muslim Moderates Must Speak Out," "More Roles for Women in Aceh, Please!", and "Discriminate Me, Acehnese Male, Please!" in Indonesian in his book, "9 langkah memajukan diri & membangun Aceh baru: rakyat ditengah arus globalisasi," (Banda Aceh: Aceh People's Forum, 2007).

than pro-active position (i.e., "we will re-consider the implementation of Syariah Law").[28] While GAM nationalist-secular political rhetoric may have attracted and in some sectors inspired the people's imagination during the armed conflict, in post-tsunami, (former) conflict Aceh, they somehow have not been able to provide answers to the spiritual soul-searching and "reconstruction of the psyche" in the transition from armed conflict to civilian life. As they themselves continue to face psycho-social accumulated trauma and other related re-integration problems, the space of reflective meditation was somehow or other left to the spiritual leaders and religious elite to fill. GAM is now left in the reactive position of how to respond to an increasingly expanding religious bureaucracy, armed with central state apparatuses and resources from Jakarta, limiting and immobilizing some of the most critical voices of opposition in Aceh.

According to several interviews conducted with INGO advisers, many of the international "experts" coming primarily (although not all) from a secularist background have been unable and unwilling to engage with religion and Islam. A few intellectually subtle international advisors I interviewed, who have been living and working in Indonesia for a long time, and who have in-depth knowledge of Islamic cultures, explained the paradoxes and contradictions by pointing to the inability of INGOs to engage in a critical dialogue with religion in general, and Islam in particular. According to one European adviser, many (though not all) of the internationals in Aceh are coming from primarily secular backgrounds (read: separation of church and state, religion and government), who see religion as "embarrassing", and either tend to hide or pretend not to have any religious background.[29] They tend to push it aside to the realm of the private and when confronted with the everyday reality of the publicness of piety in Aceh, are unable and/or unwil-

---

[28] Comments attributed to Irwandi Yusuf, the current Governor of Aceh.
[29] Conversation with Klaus Scheiner, Conflict and Islam adviser, GTZ, Banda Aceh, January 2007.

ling to grapple with religion at that very public level. Thus, the lack of rigorous engagement and the use of the "cultural relativist", or the Islamic apologist excuse, to avoid any further engagement or possible critical clashes with Islam in Aceh. With the exception of some INGOs such as Muslim Aid, Turkish Crescent, IDLO, and others who have programs that work collaboratively with the Acehnese ulama and other religious authorities, the majority of development organizations are fairly secular.

On the one hand, you have former GAM and Islamist religious leaders romanticizing and repetitively citing the historical examples of Tjoet Nyak Dhien, Malahayati, Cut Meutia, Teungku Fakinah, and their roles in Acehnese history and the independence struggle against the Dutch, and yet in contemporary Aceh under a Syariah Law regime, it is increasingly difficult for Acehnese women to participate in the public domain with new spatial constraints and limitations on roles and behaviours that women should "be". How does one explain, this paradox of actively restraining and constraining one-half of the population on the one hand, and yet on the other hand producing a lot of public rhetoric on historical "women warriors" and on women's participation as a pre-condition for democratization, reconstruction, and nation-building?

Reconstruction INGOs are not the only ones who have a blind spot in understanding Islamic cultures. Ruth McVey, an eminent political scientist working on Southeast Asia, eloquently argued 20 years ago in a critical survey of the field that: "politics is not the essence of religion."[30] McVey, who has written extensively on Indonesia, presents an excellent critique of the secular research, investigator, and/or scholar's inability to comprehend Islam as a religion: "Religious ideas and beliefs are given no independent status but are subsumed under categories

---

[30] R. McVey, "Islam Explained", *Pacific Affairs* 54, no. (1981), pp. 260-87. A similar argument was made in a recent book, see Robert W Hefner and Patricia Horvatich (eds.), *Islam in an Era of Nation-States: Politics and Religious Renewal in Muslim Southeast Asia* (Honolulu: University of Hawai Press, 1997).

such as 'politics' or 'psychology' for which social scientists already have a comprehensive and comprehensible analytical vocabulary."[31] With the exception of some excellent feminist, historical, and anthropological studies that look at Islam as a socioeconomic, religious, and cultural practice, the majority of political analyses of Islam, especially in Southeast Asia, tend to reduce it to politics.[32] Thus, McVey's argument continue to be relevant today. She writes,

> the attitude of many western social scientists for whom the tenets of a faith are unimportant except as they reflect political, economic, or other such worldly behaviour. Religion thus appears as a verbalization of certain psychological and social conditions, and it is unnecessary to consider the debates within it as arguments serious in themselves. Thus one can – and Peacock does – write an account of the relationship of Islamic modernism to the spirit of capitalism without once mentioning the question of riba (usury, unlawful interest), though this is religiously central to the issue and greatly exercised by Muhammadiyah and other Muslim thinking. The view that religion is a symptom of something else is particularly tempting for students of politics, who can conveniently take the historic identification of Indonesian religious groups with political organizations to mean that the struggle for political power is religion's primary concern.[33]

McVey's analysis above may be a useful comment in explaining why this paradox of numerous NGOs working on women's and gender

---

[31] McVey, *op. cit.*, p. 282.

[32] For a trans-historical, trans-cultural literature review of how Political Scientists, Anthropologists, Economists, Development Studies, Legal Theorists and other disciplinary fields, cover "Islam" in their scholarship, working with at least 4,000 authors throughout the world. See for example, Suad Joseph et al. (eds.), *Encyclopedia of Women and Islamic Cultures, Volume I: Methodologies, Paradigms, and Sources* (Leiden and Boston: Brill, 2003).

[33] McVey, *op. cit.*, p. 282.

issues on the one hand, and on the other hand the contradictory increase in the physical, embodied de-limiting and restriction on women's movements (at the personal, social, and body-politic level) – through control of their dress, freedom of expression of ideas, monitoring and evaluation of their sexualities and moralities.

Merle Ricklefs, another scholar of Islam in Southeast Asia insists that "if you are going to study religious phenomena you must first of all be prepared to consider the possibility that people actually believe what they say they believe. In the study of Politics, too often belief/faith is seen only in *instrumental* terms."[34] Having said this, it is also important however, to exercise what Bryan Turner offers as "critical recognition theory" – "…opportunities for reflection, dialogue, criticism…recognition involves recognition of the other, but it does not necessarily require an acceptance of their values in *toto*."[35] "Anthropologists and sociologists have generally rushed to the defence of Islam, often to the defence of fundamentalism, because they wish to avoid any accusation of racism or Islamophobia. The work of Akbar Ahmed might be taken as a characteristic illustration of a generalized defence of Islam, a sort of anthropological apologia."[36]

How does one explain the paradoxical situation of, on the one hand, a lot of funding for "gender mainstreaming", "gender justice", and "gender equality", and on the other hand Acehnese women's continuing, if not increasing marginalization in political life and the public sphere, and the tremendous challenges and obstacles they face in terms of access to justice (as in the case of victims of human rights violations during armed conflict)?[37] Some observers argue that donor money on

[34] Merle Ricklefs, personal communication, 2003.
[35] Bryan S Turner, "Classical Sociology: on cosmopolitanism, critical recognition theory and Islam," Asia Research Institute Working Paper Series. No. 39 (April 2005), p. 15.
[36] *Ibid.*, p. 16.
[37] See for example, Samsidar, (ed.), "Sebagai Korban, Juga Survivor: Rangkaian Pengalaman dan Suara Perempuan Pengungsi Terhadap Kekerasan dan Diskriminasi," (Aceh and Jakarta: Komnas Perempuan, 2006). See also urgent actions by Amnesty

"gender" could be more critically used towards something like Bryan Turner's "critical recognition theory" – of productive engagement with Islam, rather than side-stepping, if not avoidance. There are some exceptions: for example, the local NGO MISPI has a program to "develop the capacity" of women ulama; other INGOs are "capacity building" the Women's Studies Program in IAIN-Ar-Raniri. More generally, however, one gets the impression that even the most established and well-funded INGOs on gender lack visionary politics beyond so-called "gender mainstreaming". Suffice it to say that the lack of a critical self-reflection about their weaknesses and limitations is a serious problem amongst INGOs in Aceh working in the "gender sector."

There has been much criticism of the reconstruction and rehabilitation process, in particular from outspoken women in Acehnese civil society and women's organizations such as LBH APIK-Lhokseumawe, RPUK, Flower Aceh, Mispi. One of the claims is that displaced persons (women in particular) are not consulted in a participatory process and are not allowed to participate more actively or have

---

International and monthly reports by Peace Brigades International (PBI), Human Rights Watch, and TAPOL. See also, Komisi Nasional Hak Asasi Manusia Indonesia (Komnas Ham), "Laporan Tim Ad Hoc Aceh," (Jakarta: Komnas Ham, 2004), p. 84. The research for this report was led by M. M. Billah of Komnas Ham, based on primary and secondary sources, and interviews with more than one hundred eyewitnesses and victims throughout villages in five districts and six cities in Aceh. The sources also include interviews with key informants and discussions with military officials and authorities in Aceh. Field research was conducted by members of the commission and Komnas Ham staff, including local staff in Banda Aceh, and two posts in Bireuen and Lhokseumawe, throughout six months, involving ten field visits by the Commissioner, members of the commission, Komnas Ham staff, and several volunteers to assist the team. The public's perception of Komnas Ham and this report is summarized in the following comment from several people I interviewed: "Komnas Ham itu seperti harimau kehilangan gigi." [Komnas Ham is like a tiger (or other potentially powerful animal) which has been de-fanged/ or whose teeth had all been taken out.]

strong leverage in determining the reconstruction of their own houses and rehabilitation of their lives and livelihoods. Local Acehnese women observe that those supposedly in-charge of defining policies are either newly arrived, or not very concerned: for example, several of the INGO international "gender experts/advisers" are newly arrived and have yet to learn about the complex socio-political-historical context in Aceh. AMM, apparently, according to the Secretary-General of RPUK, only expressed interest in women's participation just as they were about to exit.[38]

One would think that the more progressive INGOs working on women and gender might escape the general critique of the aid industry above and the "coca-colanization" (one-size-fits-all approach) of ideas. Yet a rhetorical reading of the numerous fancy brochures and gender policies currently available in Aceh leaves one agreeing with the critical observations above. I do not have enough time and space to give a thorough analyses in this paper of various documents, but only brief critical comments on some reports.[39] For example, the Badan Rekonstruksi dan Rehabilitasi's (BRR) gender policy – "Promoting Gender Equality in the Rehabilitation and Reconstruction Process of Aceh and Nias"[40]– reads like any gender policy in any other Third World country – it could have been a gender policy in Iran, or East Timor, or Afghanistan, or Sri Lanka. There is barely anything specific to the Acehnese situation. As someone who has previously undertaken such

---

[38] See quote by Secretary General, RPUK, earlier on in the paper.

[39] A former student of mine who subsequently became a UNDP Program Officer in the "Access to Justice Program," made the following comment about these reports: "What is there to evaluate? They haven't implemented anything."

[40] BRR, "Promoting Gender Equality in the Rehabilitation and Reconstruction Process of Aceh and Nias," Policy and Strategy Paper, Deputy of Education, Health, and Women Empowerment (Aceh: BRR, 2006).

tasks (of formulating gender policies)[41], the report reads like a cut-and-paste of global governance models on "gender mainstreaming". We already know that such coca-cola models have failed elsewhere and that's why it is shocking to see them re-produced here again, in Aceh. Another example is a report entitled "The Aceh Peace Process: Involvement of Women"[42]. The report is telling in terms of its definition of "women's political involvement" and "the Aceh peace process." It is revealing in terms of its silences and the exclusion of so many Acehnese women, especially in the rural districts, who have contributed so much to political, judicial, and peace processes in Aceh – but which are completely ignored and not included in the report. Instead the report focuses on interviews with a handful of urban-Banda Aceh based women leaders and the official "peace processes". In reading several other women/gender reports similar to this one, it occurred to me that perhaps what is seriously needed is a critical re-thinking of essentialized and essentializing paradigms on women and gender, and a re-construction of the definition of "politics", and what is "political". It may even be useful to engage the question of "not partnering with the usual suspects" (i.e. only with so-called politically-correct and "beautiful people" – it may be that we have to partner with people whom we would normally not want to "hang-out" with, like the Indonesian military) and seriously abandon strategies that have a track-record of not working.

---

[41] The author has previously worked as UNDP Timor Leste Gender Adviser, Governance Unit, 2005.

[42] UNIFEM, the Centre for Community Development and Education in Aceh and the Crisis Management Initiative, "The Aceh Peace Process: Involvement of Women," Women's Voice in Aceh Reconstruction: The Second All-Acehnese Women's Congress, (2005).

Militarism, DDR, and SSR: Gendered Perspectives[43]

While it is easier to chronicle Indonesian government and TNI abuses and to portray GAM as "victims" – due to several factors, not least of which is the fact that the Indonesian government has state apparatuses to inflict state violence, and GAM has waged an armed struggle against overwhelming military, political and economic odds. It is much more difficult for Acehnese people to publicly criticize GAM as there is no 'democratic space' to do this. Critical Acehnese women articulate political subjectivities that are betwixt and between, female agency that is not often articulated publicly, for fear of being branded as "traitors to the Acehnese nation", "pro-autonomy", "anti- Syariat Islam", "secular" or much worse.[44]

The Komnas Ham and LBH Banda Aceh reports which do not pay much attention to gendered dimensions of the conflict and/or to the experiences of women, have been criticized by numerous alternative accounts from different women's organizations in Aceh, including LBH-APIK Lhokseumawe, Komnas Perempuan Special Rapporteur for Aceh, Relawan Perempuan Untuk Kemanusiaan (RPUK), Flower Aceh, MISPI, ORPAD, and several other organizations. For example, according to Samsidar:

---

[43] DDR stands for "Demobilization, Disarmament and Reintegration". SSR stands for Security Sector Reform. For gendered perspectives on "militarism as an ideology", see for example, Cynthia Enloe, "The Curious Feminist: Searching for Women in a New Age of Empire," (Berkeley, CA: University of California Press, 2004). On DDR, see "Getting it Right, Doing it Right: Gender and Disarmament, Demobilization and Reintegration," (UNIFEM, October 2004). On SSR, see for example, Geneva Centre for the Democratic Control of Armed Forces (DCAF), "Security Sector Reform and Gender," (Geneva: DCAF Publication, 2006). See also, "National Security Policy-Making and Gender"; "Defence Reform and Gender," and "Police Reform and Gender" (Geneva: Centre for the Democratic Control of the Armed Forces Publications, 2006).

[44] This is a common phenomenon in conflict and 'post-conflict' societies undergoing security, political and socio-economic transitions, but this does not justify or excuse such behaviour. This happened with Fretilin in Timor Leste, Frelimo in Mozambique, and with Fidel Castro's government in Cuba. (See again, Scott, *op. cit.*).

In the IDP camps, one of the gendered impacts of displacement is that women's access to support is usually determined by the men who dominate the IDP Committees in each camp. It is very difficult for the women to access aid addressing their specific needs, such as for example menstrual and post-partum delivery materials...pregnant women, and breast-feeding women experience malnutrition. Women who request and/or demand for the committee to address their special needs are labelled as 'women who do not know their place and selfish women who put themselves before others.' A similar resentment is expressed towards women's organizations who do provide special support to women IDPs.[45]

The report further notes that:

Data from 2003 includes 135 women experiencing violence directly related to the armed conflict...the forms of violence experienced by these women include: 17 women experiencing sexual violence, 23 women raped and 4 women subjected to sexual assault. 7 women died from being shot and killed, 11 women arrested, kidnapped, and disappeared, 22 women subjected to intimidation, destruction and looting of their property, and 50 women experiencing physical torture, five among them undergoing severe intimidation and burning of their homes. Generally, each of the victims received more than one form/pattern of violence...The reasons for the violence towards them include the victim being accused as a real or suspected member of inong balee, a sympathizer of GAM, protecting or providing food for GAM, being too close to the TNI/Polri, as a cuak (or counter-intelligence spy) for TNI/Polri. But among the 135 cases mentioned, 46 of the women did not even know, or were not given any reason as to why they were subjected to violence.[46]

---

[45] Samsidar, "Akses Perempuan Aceh pada Peradilan," in *Perjalanan Perempuan Indonesia Menghadapi Kekerasan: Konsultasi Nasional Organisasi Perempuan Indonesia dengan Pelapor Khusus PBB tentang Kekerasan terhadap Perempuan* (Jakarta, 29 July 2004), p. 72.

[46] "Akses Perempuan Aceh pada Peradilan," *op. cit.*, p.73.

A more recent report produced by Komnas Perempuan on the situation of IDPs in post-tsunami and post-Memorandum of Understanding of August 2005 is not that much more optimistic in terms of women's situation in the camps. According to the report, "Sebagai Korban, Juga Survivor: Rangkaian Pengalaman dan Suara Perempuan Pengungsi Terhadap Kekerasan dan Diskriminasi," ("As Victims, Also Survivors: Analysis of the Experiences and Voices of Displaced Women on Violence and Discrimination"),

> Discrimination, forced eviction and violence cannot be separated from the experience of IDP women in Aceh. At least 191 cases of human rights violations towards IDP women have been documented, namely 38 cases of discrimination, 7 of forced eviction and 146 of violence. These violations are perpetrated against women who are also faced with negotiating the complexity of issues related to managing aid as well as a scarcity of aid that empowers women wholistically, particularly for women who have been displaced as a result of conflict. These human rights violations are evidence that women are being alienated in all the commotion created by the reconstruction and rehabilitation `factory' operating in Aceh.[47]

Several Acehnese women used the expression: *"bermain cantik, tapi kejam"* (plays beautifully, but sharp) – referring to non-confrontational, passive opposition as a political strategy. If one is not confrontational, then one's mobility is not so constrained as one can be fairly invisible. It also ascribes a certain feminine aesthetic form, different from the way men are perceived to do politics, or what some Acehnese women disparagingly refer to as the "press conference mentality".

As in other conflict regions such as Sri Lanka, as portrayed, for example, in the poetically powerful film *"No More Tears, Sister: Anato-*

---

[47] According to Samsidar, the Special Rapporteur for Aceh, the report has been compiled based on findings of the Special Rapporteur and 20 Acehnese women IDPS who gathered data with the help and support of five mentors who worked from 19 September 2005 to 28 February 2006 to monitor 59 IDP sites located in 15 cities and districts throughout Aceh.

my of Hope and Betrayal"[48] about the life of a woman doctor who joins the Liberation Tigers of Tamil Eelam (LTTE), leaves the organization due to ideological differences and is later killed by the LTTE, Aceh women also recall and reflect on the politics of GAM as a process which involves both hope and betrayal, like two sides of the same coin. In the direct quotes below, for example, Acehnese women provide an analysis of the use of women as "human shields" (*tameng manusia*) against the TNI, the hypocritical politics of hyper-masculinity, and the cowardice of GAM:

> At the IDP camp in Bireuneun, Pidie, near the grave of Daud Beureueh, if we are allowed to speak honesty, actually many GAM members sought refuge in that IDP camp, using ordinary people as shields (*tameng*). At times when there was going to be a sweeping by TNI, the women were told to remain in the camp, but the men to move to the mosques, thinking that TNI would think twice about shooting in the mosques. But what about the women in the IDP camp? They are the ones who have to deal face-to-face with TNI interrogation. They are the ones who have to constantly answer this question: *"Mana suami mu? Suami kamu GAM?"* (Where is your husband? Your husband is GAM?) Women are used as shields. Women are put in the front line, men hide in the back. Would GAM acknowledge this? [49]

> I remember all of these clearly because we are part of all this, working with displaced persons on an everyday basis, then come the TNI kicking them around, then GAM pretending to `secure' things. TNI comes and asks: `where is your husband? where is your husband?' When in fact they know that the husband has already gone. The men always leave first, leaving the wives and children behind. When the TNI comes, it is the

---

[48] Helene Klodawsky (director) *No More Tears, Sister: Anatomy of Hope and Betrayal* (2005).
[49] Confidential interview with humanitarian aid worker, Aceh, March 2006.

women – the wives and mothers who are interrogated, made to report to TNI. Once we were in this IDP camp collecting data so that we could provide appropriate support. The TNI asked me: `where are you from? what are you doing here?" I replied: `I'm just collecting data on how many women need milk for their children and special needs that their children have.' The TNI asked us all kinds of questions. When they left and it was safe, here come GAM: `what happened just now?' I told them: `I wouldn't know. Why don't you confront the TNI yourselves. You are cowards. When the TNI was here, you disappeared. Now that they have left, then you come back here. Women have to put up with all the interrogation.[50]

As for soldiers in the villages – it wasn't a question of whether or not the women in the villages liked them. There were up to 70,000 soldiers at one point. There were those who had posts, and those who didn't. They pursued teenage women in Acehnese villages…those who were about 16, 18. Turned them into `pemuas nafsu'. Young women became involved with TNI soldiers for several reasons: 1. protection – if they had relations with a soldier their home and family felt some sense of protection. If they weren't close to the military, then they could be pestered; 2. they may need money – they could also have extra facilities. Friendship with the military could mean more facilities. Some of them became pregnant. When the soldier moved, the woman was left behind.; 3. if a soldier wanted a woman, and she didn't want – her family was threatened. So it wasn't really as if they had much choice. It's not a question of whether she likes or doesn't like.[51]

There has been much criticism of the re-integration, reconstruction and rehabilitation process in Aceh (organized by BRA and BRR), in particular from outspoken women and Acehnese civil society and human rights organizations. The general impression one has after

---

[50] Confidential interview with human rights activist and humanitarian aid worker, Aceh, March 2006.
[51] Interview with Khairani, SH, *op. cit.*

having met with and interviewed several people is that many women, while trying their best to be optimistic, see the situation as quite depressing. Many of the Acehnese women human rights activists and humanitarian aid workers interviewed for this article expressed extreme fatigue, having worked in an armed conflict situation non-stop, for many of them for more than ten years now. Post-traumatic stress disorder, as a result of several layers of accumulated trauma, with the latest being the tsunami, has taken a huge toll on the work of caring for themselves, and working in organizations supporting IDPs and victims of armed conflict. Several people commented that the tsunami wiped out several generations in one family – from grandparents, parents, to grandchildren. It also took away their brilliant and inspiring colleagues, including Syarifah Murlina, a courageous and hard-working lawyer, one of the few female lawyers in Aceh, with LBH-Medan, and also on the Board of LBH-APIK Lhokseumawe, who worked tirelessly to defend the rights of accused GAM members. She represented several prominent GAM members in their political trials and kept a well-documented archive of important human rights cases: now that she is gone, and these archives are gone, it makes it even more difficult to uncover voices and experiences of female subaltern.

Leading up to the formation of political parties for the 2009 elections, political women in Aceh are dismayed that most of the GAM political leaders are not committed to increasing more women's participation, some even actively opposing the 30% quota by saying that: "if we had that quota, we will never be able to form a party, as we won't be able to find 30% intelligent and qualified women." Khairani, SH argues: "Of course there are lots of qualified women. I'll show you – I'll put them on your list." Ironically, despite the long tradition of strong Acehnese women, some men are now saying that women's emancipation is a "foreign concept". This is exacerbated by the current overemphasis on suppressing women – their political voices, sexuality, mobility -- in the name of Syariat Islam. Women's groups also expressed anxiety about the fact that "there is currently no opposition" (to GAM leadership) in Aceh." Many people expressed their serious concern with GAM and SIRA holding state power (from Governor, Vice-Governor, Bupati, Wakil Bupati, to the geuchicks or village leaders) and no opposition movement in "civil society" or elsewhere making them accountable to corruption and criminality. According to several "security sector reform experts" I interviewed in Aceh, the biggest problem today is "extortion" (primarily from ex-GAM combatants), and other forms of new criminalities. The lack of an "alternative" political party is also a critical reflection that came up with conversations with women leaders, who are quite worried about what may happen in the 2009 elections, and have set-up their own all-women's party (to present an "alternative" to GAM and to national-Jakarta based, large political parties).

Joy Siapno began doing research in Aceh in 1992 and has authored a book and several articles on gender, Islam, nationalism and the politics of reconstruction in Aceh. She currently teaches Philosophy, Political Theory, and Political Economy at the Universidade Nacional Timor Leste and is also a Visiting Professor in the UNESCO Chair of Philosophy for Peace, Universitat Jaume I, Spain. She has also served as "Interim First Lady" in Timor Leste.

A Safer Sri Lanka? Technology, Security and Preparedness in Post-Tsunami Sri Lanka[1]

Vivian Choi

Abstract

This article focuses on the use of new technologies in Sri Lanka after the 2004 Indian Ocean tsunami in concert with recent shifts in the rationale of disaster management practices. This shift views disasters, both natural and human-made, as inevitable *risks* of everyday life to be mitigated by preparedness practices. This preparedness rationale has increasingly shifted disaster management towards disaster *risk* management. Technologies figure in disaster risk management as mechanisms to know more about disaster risks and their impacts. The more knowledge gathered on risks, the better they can be managed, ideally generating a continual state of preparedness, and in the case of Sri Lanka, a sense of national security and a government that appears to care for the safety of its people. In this article, I discuss several technologies new to Sri Lanka's disaster risk management practices. First, I explore Geographic Information Systems (GIS), which has gained global currency in disaster management efforts and the creation of an early warning system

[1] I would like to thank Malathi de Alwis for inviting me to be a part of this inspiring and encouraging research group, and for the patience and insight she and Lotta Hedman provided throughout the research and writing process. I owe a great deal to my friends and informants in Sri Lanka, especially Fajrutheen, Sidney, and the Bartholomeusz family out east. Also, comments from Nichola D'avella, Chris Kortright, and Jim Sykes were very helpful at various stages of writing.

in Sri Lanka. I study technologies such as GIS as mechanisms that follow the unfolding complexities of the post-tsunami context in Sri Lanka, lending to ethnographic work attuned to the politics of disaster risk management and national security. Rather than dismiss technology as an objective and rational tool of management, I illustrate how the push to acquire more information and knowledge about disasters constitute new technological governmental and humanitarian practices. Then, by way of ethnographic example, I show how institutional and "rational" preparedness practices unfold in one of the most devastated areas of the eastern coast of Sri Lanka during a tsunami scare in September 2007. I conclude by raising questions regarding the relationship between security, the on-going conflict in Sri Lanka, and post-tsunami reconstruction.

According to the *Scientific American*, not only was the 2004 Indian Ocean tsunami the worst disaster in "recorded" history, but also a disaster that illustrated it as one uniquely emblematic of 21st century technological developments[2]. Indeed, the tsunami marked a watershed in the implementation and utility of modern technology in disaster management. Just days after the earthquake-induced tsunami struck the shores of Indonesia, Sri Lanka, India and even parts of Africa, many an expert put technology to the grindstone, toting GPS-units to the tsunami stricken coastline, utilizing satellite images, and remote sensing all in attempts to map affected areas, illustrating, for instance, where critical infrastructure such as roads and bridges had been destroyed. Moreover, scientists and other experts were also motivated by the impetus to better understand the behaviour and impacts of the earthquake and tsunami itself. While the tsunami far exceeded anyone's (scientist or otherwise) imagination of natural disasters and destruction, it also stimulated a new fervour in the utilization of technology to

---

[2] Eric L. Geist, Vasily V. Titov and Costas E. Synolakis, "Tsunami: Wave of Change," *Scientific American* (Digital Magazine (January 2006):1, accessed in February 2008 at http://www.sciam.com/article.cfm?id=tsunami-wave-of-change.

not only study and understand disasters, but also to aid in disaster management.[3]

Given the recent occurrences of natural catastrophes worldwide such as Hurricane Katrina, the earthquakes in China and Pakistan, the cyclone in Burma, and the difficulties and conflicts surrounding the responses to them, disaster response and management practices have drawn the sympathy of a global audience and the attention of researchers in hard and social sciences in particular. Within the social sciences, some have critiqued the neoliberal underside of disaster management practices[4], while others have scrutinized humanitarian aid efforts[5]. These efforts revolve around concern for those who were affected by the tsunami. What have been people's experiences and how is the disaster being managed institutionally? This article, like others in this special issue, is concerned with how reconstruction has unfolded in the last 4 years. Particularly in Sri Lanka, how might the heightened intensity of the warfare between the government of Sri Lanka and the Liberation Tigers of Tamil Eelam (LTTE) be entangled with processes of post-tsunami reconstruction and disaster management?

As Anthony Oliver-Smith relays, disasters are notoriously difficult to study because they are multi-dimensional and processual, sweeping across every aspect of human life, impacting environments and social, political, and biological conditions, encompassing this mul-

---

[3] *Ibid.*; ESRI, "GIS and Emergency Management in Indian Ocean Earthquake/Tsunami Disaster," ESRI White Paper, (Redlands: ESRI, 2006).

[4] Cf., Naomi Klein, *Shock Doctrine: The Rise of Disaster Capitalism* (New York: Metropolitan Books, 2007). For other case studies examining disaster capitalism, including Sri Lanka, see Nandini Gunewardena and Mark Schuller, eds., *Capitalizing on Catastrophe: Neoliberal Strategies in Disaster Reconstruction* (Lanham: AltaMira Press, 2008); Jock Stirrat, "Competitive Humanitarianism: Relief and the Tsunami in Sri Lanka," *Anthropology Today* 22, no. 5 (2006): 11-16.

[5] Cf., this issue; Udan Fernando and Dorothea Hilhorst, "Everyday Practices of Humanitarian Aid: Tsunami Response in Sri Lanka," *Development in Practice* 16, no. 4 (June 2006): 292-302.; Stirrat, *op. cit.*

tidimensionality are challenging.[6] As a disaster unfolds, the manager and researcher of it must be prepared to account for its many shifting faces and articulations,[7] for as disaster traverses diverse terrains and human bodies, it acquires different meanings in different locales. Disaster is constantly articulated by the changing intersections of different forces – natural, cultural, institutional, and infrastructural. To attend to the complexities of disaster ethnographically, this article examines the post-tsunami context in Sri Lanka by focusing on the technological aspects of disaster management. Disaster management technologies have received little research attention in the social sciences, and in this article I make the case that studying the technologies utilized in disaster management can be illustrative of the politics and experiences of post-tsunami reconstruction, and in particular, linked to the intensified militaristic overtures of the Sri Lankan state against the LTTE in an effort to create a secure, united, "terror"-free Sri Lankan nation.

In Sri Lanka, the use and application of technology in post-tsunami reconstruction efforts stem from recent shifts in disaster management rationale. Amongst disaster practitioners, there has been a call to move away from disaster response and instead towards preparedness.[8] The preparedness rationale in contemporary disaster management accepts the inevitability of disasters, both man-made and natural,[9] and further, as has been pointed out by Andrew Lakoff and

---

[6] Anthony Oliver-Smith, "Theorizing Disasters: Nature, Power, and Culture," *Catastrophe and Culture: The Anthropology of Disaster*, eds., S. Hoffman and A. Oliver-Smith (Santa Fe: SAR, 2002), pp. 23-47.

[7] Kim Fortun, *Advocacy After Bhopal: Environmentalism, Disaster, and New Global Orders* (Chicago: Chicago University Press, 2001).

[8] See for instance, the United Nations International Strategy for Risk Reduction accessed on 19 December 2008 at http://www.unisdr.org/. Disaster risk management and reduction have actually been a project of the UNISDR since the 1990's, declared the decade for disaster risk reduction

[9] The man-made and natural disaster distinction is a tenuous one. There is a great deal of scholarship that questions the very "nature" of "natural disasters." This is not to deny the natural component of natural disasters – instead it is to point out that natural processes are never divorced from their social milieus. In the United States,

Steven Collier[10], views impending disasters as potential risks and therefore security threats. Hence, preparedness efforts and programs are de-

---

Hurricane Katrina served as a devastatingly poignant reminder of how socio-political relations and history are etched into natural landscapes and environments. As the levee-breaking hurricane flood waters receded, the aftermath of Katrina revealed deeply entrenched social inequalities articulated namely through the categories of race and class. Despite the notion that nature does not discriminate, it can impact lives and conditions of living disproportionately, based on prior conditions of social relations. In fact, the attribute of "natural" can serve as an "ideological camouflage for the social (and therefore preventable) dimensions of such disasters." See Neil Smith, "There's No Such Thing as Natural Disaster," *Understanding Katrina* (SSRC, 2006). Indeed, when confronting the aftermath of disasters, it is easy to sweep loss and destruction under the unpredictable and cruel rug of nature. Rather, social conditions of inequality and marginalization – often referred to as "vulnerability" – play into the experiences and impacts of a disaster. As recent research on Hurricane Katrina points out, the natural disaster could equally be considered a "human disaster" given the ensuing mistreatment and displacement of those affected – especially poor African-Americans – by Hurricane Katrina (cf., Adeline Masquelier, "Why Katrina Victims Aren't *Refugees*: Musings on a 'Dirty' Word," *American Anthropologist* 108, no. 4 (2006): 735-743.) Of course, nature as an ideological camouflage also refers to an ideological notion of nature. To serve the purposes of an ideological crutch on which to rest social woes, nature must be something external to humanity, an unpredictable "out there." This revelatory quality of disasters illustrates the tendency to separate nature from culture while also revealing the intimate relationship and mutual constitution of both nature and culture (Shannon Dawdy, "The Taphonomy of Disaster and the (Re)Formation of New Orleans," *American Anthropologist*, 108, no. 4 (2006); 719-730; Anthony Oliver-Smith, "Theorizing Disaster: Nature, Power, and Culture," *Catastrophe and Culture: The Anthropology of Disaster*, eds. Anthony Oliver-Smith and Susanna Hoffman (Santa Fe: SAR, 2002), 23-47.). Throwing into sharp relief ideological notions about human-environment relations, disasters challenge our presuppositions, "the very horizon" of our understanding of and the meanings we give to "nature" (Slavoj Zizek, *Looking Awry: An Introduction to Jacques Lacan through Popular Culture* (Cambridge: MIT, 1991).

[10] Andrew Lakoff, "The Generic Biothreat, Or, How We Became Unprepared," *Cultural Anthropology* 23, no. 3 (2008): 399-428; Andrew Lakoff, "Preparing for the Next Emergency," *Public Culture* 19, no. 2 (2007): pp. 247-271; Andrew Lakoff and Stephen Collier, "Vital Systems Security," Discussion Paper, Anthropology for the Contemporary (February 2006).

signed to manage risks, and what was before understood as disaster management is now seen as disaster risk management. Concomitant with this shift in rationale, then, is an impetus towards "knowledge," which requires the culling of data and information regarding the disaster itself and its impacts. The idea is that the more knowledge one has about a risk, the better it can be managed, ideally generating a continual state of preparedness, and in the case of Sri Lanka, a sense of national security and a government that appears to care for the safety of its people.

One of the main ways that such knowledge is produced within this broader shift towards risk management in Sri Lanka is through the introduction and implementation of new technologies. This article will focus on these newer technologies of preparedness, highlighting the shift in disaster management that is currently underway. In particular, I am interested in Geographic Information Systems (GIS), which has gained global currency as an essential technology in disaster management efforts[11] and the creation of an early warning system in Sri Lanka. I approach the study of technologies such as GIS as a way to understand the unfolding complexities of the post-tsunami context in Sri Lanka.[12] Rather than dismiss technology as merely an objective and rational tool of management, I study how the recent push to acquire more information and knowledge are part and parcel to new technological governmental and humanitarian practices. The application of technology requires the formation of new knowledges and the collection of data information. I will also discuss the recent development of

---

[11] ESRI, *op. cit.*; interviews at the Disaster Management Centre and notes from GIS and disaster management workshops in Colombo 2007.

[12] Kim Fortun, "Post-structuralism, Technoscience and the Promise of Public Anthropology," *India Review 5,* no. 3 (December 2006): 294-317; Geoffrey Bowker and Susan Leigh Starr, *Sorting Things Out: Classification and Its Consequences* (Cambridge: MIT Press, 1999); Geoffrey Bowker, *Memory Practices in the Sciences* (Cambridge: MIT Press, 2005); Annemarie Mol and John Law, "Complexities: An Introduction," *Complexities: Social Studies of Knowledge Practices*, eds., Annemarie Mol and John Law (Durham: Duke University Press, 2002), pp. 1-22.

an early warning system in Sri Lanka. By way of ethnographic example, I illustrate the possibility of co-existing practices of preparedness and disaster response – that is, how institutional and "rational" preparedness practices towards greater security played out in one of the most devastated areas of the eastern coast of Sri Lanka, when there was a tsunami scare in September 2007. In doing so, I illustrate that studying technologies avail themselves to productive ethnographic examination; the structure of the technologies themselves lends to dynamic ethnography attuned to the politics of disaster and disaster risk management and national security in Sri Lanka.

"A Road Map to a Safer Sri Lanka":
A Preparedness Rationale

On December 26, 2004, Sri Lanka was caught off guard. Without warning, the waters receded and returned with a mighty vengeance. The unexpected tragedy elicited an "if only we were prepared" form of regrettable sadness. If only there was a warning system, if only people had known that a retreating sea signals danger. Recent disaster management pushes have turned these if's into when's. The rationale of the disaster management cycle featured in Figure 1[13] illustrates recent shifts in disaster management rationale – a rationale that expects disaster. This rationale of preparedness considers disasters as potential risks, wherein disaster management becomes risk management. In his analysis of aftermath of the September 11th tragedies in New York and Hurricane Katrina, Andrew Lakoff stresses that framing disasters as risks, whether natural or human-made, categorizes them as forms of threat, which in turn, brings them all under a common national security field. Under the purview of preparedness, disaster mitigation solicits new technical band-aids: warning systems, infrastructure management, and an overall increase in government management and

---

[13] Accessed on 19 December 2008 at
http://www.dmc.gov.lk/Phases_%20of_%20the_Disaster.html.

Figure 1: Disaster Management Cycle, Sri Lanka's National Disaster Management Centre website

control.[14] It also enlists the work of various different actors such as scientists, urban planners, humanitarian aid, and governmental control. The impetus of preparedness efforts is to invoke a continual state of readiness and maximum security of state territory. As Lakoff writes: "This treatment involves first, tracking the occurrence of such events over time across a population; and second, applying probabilistic techniques to gauge the likelihood of a given event occurring over a certain period of time ... what had been exceptional events that disrupted the normal order become predictable occurrences."[15] The abnormal, in this sense, is made normal; "disaster" becomes naturalized, indomitable.

---

[14] Ulrich Beck, *Risk Society: Towards a New Modernity* (Thousand Oaks: Sage, 1992); Craig Calhoun, "A World of Emergencies: Fear, Intervention and the Limits of Cosmopolitan Order," Annual Sorokin Lecture, University of Saskatchewan (4 March 2004).

[15] Lakoff, 2007, *op. cit.*

Following the tsunami, the Sri Lankan government assembled the Sri Lankan Parliamentary Committee on Natural Disasters, whose mandate was to assess the level of preparedness towards such unexpected disasters. The culmination of this Committee was the Disaster Management Act No. 13 in May 2005:

> This Act provides for a framework for disaster risk management in Sri Lanka and addresses disaster management (DM) holistically, leading to a policy shift from response-based mechanisms to a proactive approach toward DRM (Disaster Risk Management).[16]

As outlined in the "Roadmap for a Safer Sri Lanka", "safer" primarily entails a reduction of risk, in which assessments of vulnerability, hazard and risk assessment are key in creating a state of preparedness for whatever type of disaster comes. This shift towards preparedness as opposed to responsiveness is being overseen by the newly created National Disaster Management Centre undertaking the express "… mission to create a culture of safety to reduce the vulnerability of the population to natural hazardous events in the future."[17] While certain specialized disaster agencies were already in existence in Sri Lanka, there was no legal framework for disaster management and thus no holistic mechanism by which to coordinate it. What is notable is that the pivotal interventions, the vulnerability, hazard and risk assessment thematic section of the "roadmap" require a shift towards knowledge-based systems and data collection. One of the main activities of the Disaster Management Centre is a GIS-based disaster risk management system, which includes disaster zonation mapping, and an

---

[16] Final Report of Sri Lanka Parliament Select Committee on Natural Disasters, May 2005 accessed on 19 December 2008 at
http://www.srilankanparliamentonnaturaldisasters.org/
Key%20Recommendations.htm.
[17] *Ibid.*

overall coordination and construction of a disaster database to be updated and readily available for public use and mapping.[18]

A Short Primer on
Geographic Information Systems (GIS)

Globally heralded for its ability to visualize and represent disaster,
GIS cannot only map spaces of disasters but also can align other relevant
features, conditions, and events with those geographical spaces. For example, immediately after the tsunami in Sri Lanka, GIS maps were utilized to view washed out roads and bridges, to identify usable routes by
which to deliver aid resources, and to show locations of existing Internally Displaced Persons (IDPs) camps. In more recent efforts, now that
Sri Lanka is engaged in a phase of reconstruction (although this phase,
it seems, will slowly fade out), GIS technology continues to play a substantive role in post tsunami-related activities: mapping out land-use
policies and community resettlement projects; coordinating the presence of humanitarian aid and the geographic locations of their projects;
and early warning/disaster preparation systems. This indispensability of
GIS technology in disaster contexts has spurred global initiatives on
disaster reduction and recovery. This is especially the case in Sri Lanka,
where government officials, university professors and international aid
and humanitarian organizations have convened at meetings and workshops to assess the values and progress of GIS in disaster management.
Specifically, because GIS functions as a data consolidator, wherein sources
of information relevant to the disaster situation can be culled into a central
database[19] there has been a recent emphasis on technological development
and standardization of data to foster data interoperability and sharing

---

[18] On a somewhat related note, the World Bank is now implementing programs to
encourage Sri Lanka's knowledge economy. "Harnessing knowledge" includes creating an "adequate and modern" technology infrastructure in order to aid in economic
"productivity and development." For further reference: See, "Building Sri Lanka's
Knowledge Economy," A World Bank Report, March 2008.

[19] ESRI, *op. cit.*; R.W. Greene, *Confronting Catastrophe: A GIS Handbook* (Redlands,
CA: ESRI, 2002).

and communication.[20] When a GIS map is produced, the information that is presented is actually layers of "information." Once a GIS database is established, the user may create a map using the information in the database.

Among GIS users and practitioners, there exists a broader history of debate on defining GIS. My research employs a definition given by geographer Nicholas Chrisman[21]:

Geographic Information System (GIS): The organized activity by which people: measure aspects of geographic phenomena and processes;

represent these measurements, usually in the form of a computer database, to emphasize spatial themes, entities, and relationships;

operate upon these representations to produce more measurements and to discover new relationships by integrating disparate sources; and

transform these representations to conform to other frameworks of entities and relationships.

Chrisman's definition attempts to capture the processual, shifting, and experimental aspects of building and using a GIS, to open up the GIS "black box" and illustrate its profoundly socially contextual nature, rather than to understand it as simply an input-output tool with inert and objective data. Indeed, the technical components of GIS do not operate in a vacuum. As Chrisman elaborates: "measurements, representations and transformations all serve the goals of institutions, and these, in turn, serve larger social goals. But the information is not sim-

---

[20] ESRI, *op. cit.*
[21] Nicholas Chrisman, "What does 'GIS?' Mean?" *Transactions in GIS* 3, no. 2 (1999): 175-186.

ply a passive player, responding dutifully to social demands. The availability of information shapes social expectations and the cultural expectations of professions and disciplines shape the choices of measurement and representation."[22] As other practitioners of GIS highlight, building and implementing a GIS takes much labour, coordination, and the negotiation of numerous actors and stakeholders[23] and is an exercise in creating uncertain spaces and boundary objects.[24]

Understanding the various arguments and practices of GIS is to learn how it is built and how it is employed, specifically in the context of post-tsunami reconstruction and disaster risk management efforts in Sri Lanka. It is precisely because GIS require the creating, culling, co-ordination, inputting of data, all of which precede the visual articulation that emerges as the final product, that I employ it as a tool by which to examine the complex networks, relations and spaces that are being articulated and produced in post-tsunami reconstruction and disaster preparedness practices. While the representation and representational effects of maps are certainly worthy of critical attention,[25] I focus on how methodological pursuits of disaster information and mapping inhere in the form of the GIS itself[26]. GIS can both serve as both form and object of research and guide my examination of the ways in which the tsunami and "disaster" is translated through various

---

[22] *Ibid.*

[23] Barbara Poore, "The Open Black Box: The Role of the End-User in GIS Integration," *The Canadian Geographer* 47, no. 1 (2003): 62-74.

[24] Nadine Schuurman, "The Ghost in the Machine: Spatial Data, Information and Knowledge in GIS" *The Canadian Geographer* 47, no. 1 (2003): 1-4.; F. Harvey and N. Chrisman, "Boundary objects and the social construction of GIS technology," *Environment and Planning A* 30, no. 9 (1998): 1683-1694.

[25] For example: Pradeep Jeganathan, "eelam.com: Place, Nation, and Imagi-Nation in Cyberspace," *Public Culture* 10, no. 3 (1998): 515-528; and Thongchai Winichakul, *Siam Mapped: A History of the Geo-Body of a Nation* (Manoa: University of Hawaii Press, 1994).

[26] Annelise Riles, "Infinity within the Brackets," *American Ethnologist* 25, no. 3 (1998): 378-398; and Annelise Riles, *The Network Inside Out* (Ann Arbor: University of Michigan Press, 2000).

engines and institutions of data collection and mapping. Studying the role of GIS in disaster management, then, is to also become privy to the politics that surround its construction and application, and the politics surrounding post-tsunami reconstruction.

GIS as Ethnographic Tool

As such, I see in the structure of GIS an instrument and artefact of disaster and risk management, a network of relations. What I wish to offer, then, is a way to figure GIS technology into the ethnographic process itself that does not necessarily pit it simply as an "objective" form of calculative power. To take the technology seriously is not to dismiss it wholesale as a dangerous god-trick, an objective rationality of state power, neither is it to accept the technology as a benign representation and organization of the lived world. The realm of knowledge and data production is a space worthy of ethnographic attention where knowledge workers and data technicians also undertake the burden of building a "safer" Sri Lanka.[27] This knowledge and self-awareness thread laterally across institutions as well as out on the ground to those living in designated zones of disaster-affected. The current regime of technological know-how and disaster risk management produce and materialize new forms of life and subjectivities; this includes these data and mapping managers and makers as well as those who have been more visibly or materially affected by the tsunami.

Because the development of these new technologies of disaster management require so much labour, knowledge, and collaboration, certain ethnographic questions emerge. For instance, how are contingently created and varying contexts and experiences made commensurable? Through what politics and powers are data and knowledge sieved

---

[27] Specifically I am inspired by the notion of "ecology" as an environment and collaboration of experts and expertise: Aihwa Ong, "Ecologies of Expertise: Assembling Flows, Managing Citizenship," in *Global Assemblages: Technology, Politics, and Ethics as Anthropological Problems*, eds., Aihwa Ong and Stephen J. Collier (Malden: Blackwell Publishers, 2005).

through the perspectival notions of human rights, the nation, and the global? How are subjectivities and existences articulated in relation to forms and configurations of state power? If data must be created it must also be managed.[28] With these questions in mind, I first highlight disaster data collection and management in institutional settings.[29] I will then follow by discussing the implementation of a multi-hazard warning system in the eastern coast of Sri Lanka, the worst tsunami-affected region.

Who What Where:
Coordination among Humanitarian Agencies

Coordination quandaries plagued Sri Lanka in the immediate aftermath of the tsunami. Although no stranger to the presence of international aid and assistance due to existing humanitarian crises related to its protracted civil conflict, after the tsunami Sri Lanka found itself awash with a "second wave" of humanitarian aid. That various international aid organizations and workers, scientists, and even volunteer tourists inundated the distraught island in the aftermath of the tsunami, introduced yet another dimension of disorder – the outpouring of support was difficult to coordinate. The Sri Lankan government, receptive of the assistance, nevertheless had difficulty managing who was doing what, when and where in the country. As Jock Stirrat observed in southern coastal tsunami-affected areas, lack of coordination and the prolific presence of aid and aid organizations resulted in

---

[28] Geoffrey Bowker, *Memory Practices in the Sciences* (Cambridge: MIT Press, 2005); Marilyn Strathern, "A Community of Critics: Thoughts on New Knowledge," *Journal of the Royal Anthropological Institute* 12 (2006): 191-209.

[29] Recent anthropological works have proposed the value in working with institutions and experts. Cf. Lakoff and Collier, *op. cit.*; Douglas Holmes and George Marcus, "Cultures of Expertise and the Management of Globalization: Towards a Re-Functioning of Ethnography," in *Global Assemblages: Technology, Politics, and Ethics as Anthropological Problems,* eds., Aihwa Ong and Stephen J. Collier (Malden: Blackwell Publishers, 2005), pp. 235-252; Jennifer Hyndman, *Managing Displacement: Refugees and the Politics of Humanitarianism* (Minneapolis: University of Minnesota Press, 2000).

counter-productive competitiveness, wherein international aid organizations found themselves "carving out" their own territories for assistance, ever mindful of the boundaries forged by other organizations.[30]

In order to manage a disaster efficiently and appropriately, information must be managed as well. According to the UN Office for the Coordination of Humanitarian Affairs (UNOCHA):

> The key information required to assess and ensure that humanitarian needs are met in any emergency/disaster is, to know which organizations (Who) are carrying out what activities (What) in which locations (Where) which is also universally referred to as the 3W (Who does What Where)[31]

As such, a database for the "3W" is produced and "universally" agreed to be the "most important priority" for any coordination activity in disaster management. In this effort, a UNOCHA field office distributes a 2-page standard data collection sheet, detailing specifics of the project (Who What Where) to be filled out by the participating agency. The sheets are then checked by the field office and then sent to the main UNOCHA in Colombo and then added to the 3W GIS database, where it will then be mapped and published for public viewing at the UNOCHA Humanitarian Portal.[32] The Humanitarian Portal also includes links to all maps that are being made, meeting schedules and detailed meeting notes, weekly reports by the Inter-Agency Standing Committee of Sri Lanka (IASC in UN parlance).

Despite this effort of coordination and seeming transparency, there is no assessment of the effort itself. While it is possible to get a sense of who is doing what and where, it is difficult to imagine who else might be doing what and where, as the creation of the database re-

---

[30] Stirrat, *op. cit.*

[31] UNOCHA, "Who What Where," accessed on 19 December 2008 at http://www.humanitarianinfo.org/sriLanka_hpsl/ whowhatwhere.aspx.

[32] See, UNOCHA, "Humanitarian Portal – Sri Lanka," accessed on 19 December 2008 at http://www.humanitarianinfo.org/ sriLanka_hpsl/.

lies on the responsibility of active organizations to report on their projects. Or alternatively who is not doing anything where. My intent here is not to critique humanitarian operations or their ideologies. Rather, it is to recognize that transparency is not always transparent and consider humanitarianism not as an absolute value, but as an array of embodied, situated practices that stem from the humanitarian impetus to relieve suffering. It is also to recognize that out even out of humanitarian concern, chaotic and disorganized responses and interventions to complex emergencies can lead to the recurrence or exacerbation of crises rather than development out of that cycle. Information does not always lend to more "knowledge." Humanitarian action occurs, in other words, in an "untidy, thoroughly implicating, 'second best world.'"[33]

If Only We Knew:
Making Data, Becoming Knowledgeable
There is a joke that supposedly "everyone" knows or has heard, in varying versions, about the tsunami in Sri Lanka. It goes something like this:

> The phone rings at a government office in Colombo. An administrator picks up:

> "Hello?"

> "Yes, this is the earthquake monitoring centre. We wanted to inform you that a T-tsunami will be arriving in approximately 2 hours."

> "A T. tsunami? Yes yes, we'll make the necessary arrangements to receive T. Tsunami."

---

[33] Originally quoted by Fiona Terry, *Condemned to Repeat? The Paradoxes of Humanitarian Action* (Ithaca: Cornell University Press, 2002); taken from Peter Redfield, "Doctors, Borders, Ethics of Crisis," *Cultural Anthropology* 20, no. 3 (2005): 328-361.

And so it goes that the administrator sends a car to the airport to pick up a Mr. T. Tsunami and the car waits for some time. But alas, it seems a Mr. T. Tsunami is not coming – at least not by plane. Actually, by this time, the t-tsunami had already arrived – as everyone knew – by sea.

At first, I listened to this story in disbelief. Each time, much laughter followed its telling. I realised it was a joke told with much self-deprecation and irony. Often times, I have heard and am told that people substitute the administrator receiving the phone call with names of politicians they think incompetent. While it is admirable that people can have a sense of humour regarding such shocking destruction and devastation, this joke also highlights the recognition of a lack of knowledge that many people had at the time, including government officials and coastal dwellers, about such catastrophically powerful and deadly phenomena such as tsunamis. Furthermore, like many other jokes which do the rounds in Sri Lanka, it is subversive. It provides a critique of Sri Lankan state power, peopled by incompetents; a government upon which little confidence is bestowed when it comes to security and safety. [34]

In the wake of the 2004 tsunami, "If only we knew" has become a familiar lament. And this lack has ushered in new forms and spaces of knowledge- and security-making in Sri Lanka. [35] The creation

---

[34]This joke also echoes what is now considered to be an egregious act of negligence on the part of those manning the Seismic Monitoring Centre at the University of Peradeniya who were not at their desks to receive intimation of an impending tsunami, from the Pacific Tsunami Warning Centre in Hawaii, as it was a public holiday in Sri Lanka (I am grateful to Malathi de Alwis for providing this additional overlay to this joke).

[35] Kim Fortun has discussed how "information is not only of substantive value (valuable because of its potential truth content) but also because of what can be called semiotic value. Any piece of information – even if partial or lacking verification – can draw people into processes of inquiry, driven by recognition of potential but unrealized information density, of interests undergirding information gaps, and of varied ways information, even if questionable, can be used – for comparisons across space and time, for example. Information thus creates *capacity* to understand and respond

of the National Disaster Management Centre in 2005 in Sri Lanka illustrates how information collection and knowledge acquisition have become a priority in national disaster risk management and securitization efforts. In August 2006, I met with the director of the newly government-created National Disaster Management Centre. While there, I became privy to the plan to implement a new disaster monitoring system and database called DISENVENTAR. The basic design of the system is to catalogue all disasters occurring throughout the country on a monthly basis. Through the cooperation of local administrative sections of the government (District Secretariats and *Grama Niladharis*), disasters are to be reported monthly and sent to the Centre. Disasters include floods, droughts, landslides, and economic costs are to be documented as well. For instance, some of the questions that appear on each reporting form include: What type of disaster? What caused the disaster? How many people were affected? For how many days? What were the economic costs? In addition, old newspapers and reports were being mined to tabulate the occurrences of disasters in the past, starting from 1976. The purpose is to compile a running tabulation of disaster occurrences, a monitoring system that might be able to establish some would be in a position to better predict the next disaster, and more importantly, on account of compiling the costs and effects of disasters, would be better prepared. The key idea was the acceptance of the What is also of interest, is that in the disaster inventory system, a new category was established following a spate of alleged LTTE bombings of Colombo the capital city of Sri Lanka: Terrorist Attacks.[36]

---

to problems, routine and catastrophic." (Kim Fortun, "Environmental Right-to-Know and the Transmutations of Law," ed. Austin Sarat (Minneapolis: University of Minnesota Press, forthcoming 2009). My research views information technologies as ethnographic opportunities and openings to study disaster and disaster risk management in Sri Lanka.

[36] Following this visit, DESINVENTAR has since migrated and become a major project for the Disaster Management Centre (DMC) in collaboration with the United Nations Development Program (UNDP) under the Ministry of Disaster Management and Human Rights. There is a focus only on natural disasters and

When terrorist attacks and natural disasters become abstracted, and in turn commensurable as "disasters", this move highlights that bombings and other forms of violence are also an inevitable part of life in Sri Lanka. As the Sri Lankan government is poised to spend a record 1.82 billion dollars on its defence budget in 2009 in order to continue its military attacks on the LTTE, it follows that they must also be prepared for counter attacks and bombings. It is no surprise, then, that my follow-up visits to the Disaster Management Centre revealed that all data collected for DISINVENTAR must await the stamp of approval and validation by the Defence Ministry. Preparedness appears to be the hallmark of the contemporary security moment, where knowledge and database systems are gaining significance in the "information dimension of sovereignty."[37] This recent move by the government certainly

---

preparedness and mitigation activities and programs. As one program officer informed me, getting the information on terrorist attacks would be too difficult to coordinate with the Ministry of Defence and furthermore is not for public knowledge. The Ministry for Resettlement and Disaster Relief Services is responsible for so-called "post" disaster issues. This ministerial separation, due in part to a desire to keep in place certain management positions, also fulfills disaster issues following the model/phases of disaster risk reduction, as illustrated in Figure 1 and discussed earlier in the article. The DMC accounts for preparedness protocols, while the Ministry for Resettlement and Disaster Relief Services focuses on disaster response and rehabilitation, including the plight and resettlement of Internally Displaced Persons (IDPs). Despite the fact that DESINVENTAR no longer includes "terrorist attacks" as a disaster type and category (although I have been assured by program officials that it is certainly very easy and possible and might be integrated into the system later), the overlap between "natural disasters" and "terrorist attacks" is maintained in their afteraffects – that is, internal displacement and relief services are not separated based on their "causes." In fact, in the northern and eastern regions of Sri Lanka, the Asian Development Bank Country Director remarked, "pre-tsunami conditions in the North East [made] it difficult to demarcate tsunami- and conflict-related rebuilding needs". (See, http://www.adb.org/media/Articles/2005/ 7343_Sri_Lanka_projects/, last accessed 7 March 2009). Accordingly, the push in reconstruction and rehabilitation activities in northern and eastern Sri Lanka, was to "return to normalcy," furthermore, begging the question of what "normalcy" means in situations with persistent insecurity and trauma.

[37] Karen Litfin, "Satellite and Sovereign Knowledge: Remote Sensing of the Global

merits further attention, and I will conclude with how these actions relate to Sri Lanka's on-going and increasingly militant and violent civil war. In the meantime, the Disaster Management Centre, with support from the United Nations Development Program (UNDP), continues to implement preparedness activities. The political logic of security that undergirds the preparedness rationale not only creates a knowledgeable government in order to maintain a semblance of security, but also knowledgeable citizens, who must be ready and equipped for impending disasters. Let us see what this looks like.

One of the latest projects undertaken by the Disaster Management Centre is the creation of an early warning system in Sri Lanka. With the support of the United Nations Economic and Social Commission for Asia Pacific (UNESCAP) and the South Korean government, the Disaster Management Centre, together with the Department of Meteorology in Sri Lanka have started the process toward a Multi-Hazard National Warning System.

Plagued by the notion "if only we had known," it is believed that a warning system could have averted the extensive damage and death caused by the tsunami in 2004. Given the existing preoccupation of data collection, the implementation of an early warning system has hence been at the forefront of disaster risk management, because an early warning system requires continual observation and data collection on disasters and hazards. Beginning in 2005, a pilot project to set up disaster warning towers in three tsunami-affected locations: one in Kalmunai on the east coast (see Figure 2), one in Hikkaduwa on the south coast and one in Point Pedro in the Jaffna Peninsula (north coast). The towers in Kalmunai and Hikkaduwa were officially "opened" on the second anniversary of the tsunami. The tower in Point Pedro, due to security reasons and lack of telecommunications infrastructure is not yet "running".

Environment," in *The Greening of Sovereignty in World Politics*, ed., K. Litfin (Cambridge MA: MIT Press, 1998), pp. 207–8.

Figure 2: Disaster Warning Tower in Kalmunai, Eastern Province,
(Photo courtesy of the National Disaster Management Centre, 25 March 2008.)

The disaster warning towers are controlled by the main opera-tion centre located in Colombo, at the Disaster Management Centre. The Emergency Operations Centre is a large room, lined with several flat screen televisions (when I visited, two were set on news stations) and a larger more formidable-looking computer, the central operating system, which receives and distributes messages to the appropriate peo-ple and channels. The Centre communicates with other agencies such as the Pacific Tsunami Warning Centre, the United States Geological Survey, and the Japan Meteorological Agency to obtain and monitor the most up-to-date information on possible hazards. Open 24 hours a day, 7 days a week, the goal of the National Emergency Operations Centre is to disseminate disaster related information, and in the case of the early warning towers, send alert messages to district officials, the police and army and also remotely control the sounding of warning si-rens on the towers.

Of course, once a warning is issued, those living in endangered areas must also know the proper protocol. In this regard, disaster edu-cation and capacity building have become large parts of reconstruction efforts, where the focus is on the structuring and institutional man-agement of the technology itself. The disseminated emergency informa-tion must have an educated audience, in order to make it truly effective. Or does it?

On 13th September 2007, the Sumatran-Andaman subduction zone near Indonesia that elicited the *periya* ("big") tsunami in 2004, heated up once again, causing a record-high tremor of 8.4 on the Rich-ter scale and setting off warnings. Sri Lanka's Disaster Management Centre in Colombo received a warning message as well. In the eastern coast's Kalmunai, though the tower was officially "open" or working, the warning siren, however, did not sound. The local disaster manage-ment officer living in the village where the tower is located informed me that communication between the central office and tower was not going through. He told me that he waited to hear from the central command, to get directions as to what to do. At this point, he noticed that local villagers had started evacuating already, running inland, clogging up the main road and taking shelter in the local mosque. Apparently, Dialog™, a Sri Lankan mobile phone service provider, had already sent SMS or text messages to its user network, with the warning of another tsunami. Forty-five minutes after that text message was dispatched, the military Special Task Force arrived, encouraging an emergency evacuation. It was at this time that the district disaster management officer physically climbed up the warning tower and sounded the siren, despite the fact that most living near the shoreline had already begun the evacuation process. Fortunately, the earthquake was located just far enough South that the tsunami everyone feared would strike again did not happen.

According to the disaster management officer, the central operations centre had difficulty communicating with regional officers and warning towers because telephone lines were completely tied up. As text messages went out and fearful rumours of another tsunami spread, people began to call everyone they knew who might be in danger or to call and reassure other people that they were aware of the impending danger.[38]

"Yes, I received a call from my brother in Trincomalee that a tsunami was coming," Fathima[39] relayed to me. Her brother, a fisherman, had been listening to the radio, and had heard some news about

---

[38] In response to the confusion surrounding the 13th September tsunami scare, an interesting discussion concerning the use and reliability of SMS and mobile phone warnings in Sri Lanka has transpired: "SMS alerts during emergencies - Lessons from Sri Lanka's tsuanmi alert on 13 September 2007," ICT for Peacebuilding (ICT4Peace), 13 September 2007, accessed at http://ict4peace.wordpress.com/2007/09/13/sms-alerts-during-emergencies-lessons-from-sri-lankas-tsuanmi-alert-on-13-september-2007/; "SMS news alerts during emergencies - The experience of JNW and the tsunami warning of 13th September 2007," *Groundviews*, accessed at http://www.groundviews.org/2007/09/13/sms-news-alerts-during-emergencies-the-experience-of-jnw-and-the-tsunami-warning-of-13th-september-2007/.

[39] I have changed names to preserve the anonymity of my informants.

an earthquake in Indonesia, which immediately prompted him to call his sister, who lives in the area that was most-affected in the tsunami of 2004. Fathima's home is roughly 10 metres from the warning tower. She said she did not wait to hear it ring, but instead gathered up her family members, made some phone calls to relatives and close friends and began to spread the word to her neighbours. She said that in a matter of a half hour, nearly everyone in her village had begun to make the journey inland to the main road. She took refuge in her brother's home, just on the other side of the main road, while others, she said, headed to the local mosque, or simply waited on the main road, causing traffic jams for those attempting to drive inland.

Had the warning tower failed her because it did not sound the alarm that day? On the contrary, she assured me, the mere presence of the tower offered her a sense of security. Just knowing that it was there, to warn them in case of another tsunami was enough for them to feel that the type of devastation wrought by the Indian Ocean *"periya"* tsunami would not happen again. If the tower had existed before, Fathima told me, she believes that it would have saved uncountable lives. Mohamed, a local owner of a small *kadai* on the beach also relayed the same sentiment to me as I sipped a warm Coke near the cooling breeze of the ocean. In fact, because of the warning tower, he felt comfortable enough to re-open his shop near the beach. "At night I can sleep better," he said, looking towards the expanse of ocean in front of him. He blinked slowly, "Before I could only listen to the sound of the waves, listening for the way the waves sounded the day of the tsunami."

In providing the details and experiences of this event, my goal is not to point out the shortcomings of the early warning system. Indeed, I do agree that measures should be taken to protect those living in dangerous areas. Disaster responses and preparation should be improved. However, I do not accept these technologies wholesale either, and I seek to question the confidence that is put in technology's contribution to social engineering, especially social engineering initiated by institutional powers. I want to show that establishing and implementing techniques of preparedness are difficult and might improve through trial and error[40]. Moreover, while it appears that the government provides some form of security and semblance of control through the establishment of the early warning system, there are still other forms and networks of security and technologies animated on the ground. I use "on the ground" not as an opposition to the emergency operations centre, but rather as a location, a site among many where the rationale of disaster risk management plays out. As Andrew Lakoff and Stephen Collier suggest, it is analytically productive to ask: "which forms of collective security are in question, what kinds of expertise are being mobilized to provide security, and how the politics of security are changing?"[41]

A Safer Sri Lanka or a False Sense of Security?

It is not without some sense of irony that the roadmap to a "safer Sri Lanka" is being paved on an existing and increasingly militant context of war. Joining terrorist attacks and natural disasters into an abstract category of disaster risks the danger of accepting bombings and violence as a normal or inevitable risk of daily life – as part of a "safe" life. To be sure, "normal" life is full of dangers and risks, but the making of such a normalizing abstraction might remove the culpability of the government's investment in a full-scale war and how it affects posttsunami reconstruction and development. In a preparedness rationale, as different types of harmful events are conceived as disasters and therefore issues of national security or threat, it becomes apparent that conflict and disaster risk management in Sri Lanka are inextricably woven together. When looking at the relationship between the war and the economy in Sri Lanka, for example, Deborah Winslow and Michael Woost focus on "the shifting articulations of understanding, practice and experience that have been central in the expansion of violence in

---

[40]The Disaster Management Centre now uses satellite telecommunications, having learned from their experience in September 2007.

[41] Lakoff and Collier, *op. cit.*

Sri Lanka generally and in the war's production and reproduction."[42] In this regard, is disaster risk management as a form of security also central to the Sri Lankan government's violent pathway to peace?

Nearing the end of 2008, the Sri Lankan government's defence budget soared to record-high levels, fulfilling earlier projections of making 2008 the year for war. In September 2008, the government ordered humanitarian agencies and Non Governmental Organisations to evacuate the Northern LTTE stronghold in Killinochchi. This evacuation is due in large part to the fact that the government cannot assure the safety of these workers. If the government cannot assure the safety of international aid workers, how can the safety and needs of locals, IDPs and others still living in zones ravaged by continuous warring be guaranteed? Is this the pathway to building a "safer" Sri Lanka?

Vivian Choi is a PhD student in the Department of Sociocultural Anthropology at the University of California, Davis. She is currently engaged in field research in the Eastern Province of Sri Lanka.

---

[42] Deborah Winslow and Michael Woost, "Articulations of Economy and Ethnic Conflict in Sri Lanka," in *Economy, Culture, and Civil War in Sri Lanka*, eds., Deborah Winslow and Michael Woost (Bloomington: Indiana University Press, 2004), pp. 1-30.

From Research to Policy: The Case of Tsunami Rehabilitation in Sri Lanka

Sunil Bastian

Abstract

The objective of this paper is to identify the policy implications of the findings of the comparative project on tsunami rehabilitation carried out in Sri Lanka and Aceh. It begins with a critique of the notion of 'emergency' that dominates humanitarian organisations. It also shows how managerial tools like Guiding Principles are no answer to the complex issues faced by projects. The last section points to a number of issues raised by the research studies: the importance of constantly questioning the theoretical and methodological models used by humanitarian aid organizations; the need for a better understanding of the nature of the state and its possible role in rehabilitation; the tragic consequences which result when social organization and land tenure patterns are ignored by aid organizations; that the repetition of the mantra of community participation does not ensure equity. It concludes with the argument that for policies and projects to be successful, implementors have to make use of the existing knowledge base and must employ people who have the competence and experience to work in these societies; the institutional structures of the agencies must also be flexible to deal with all kinds of social and political complexities which are unique to each country.

The research project on which this special issue of *Domains* is based, 'Post-tsunami reconstruction in contexts of war – a grassroots study of the geo-politics of humanitarian aid in Northern and Eastern Sri Lanka and Aceh, Indonesia,' departed from the approach that dominated much of other studies on tsunami rehabilitation, in important ways. Firstly, it avoided the period of heightened interest in the tsunami, from 2005-2006, and sought to focus on longer term repercussions of rehabilitation. Secondly, it was based on eight, long-term field-level studies carried out in Sri Lanka and Indonesia. Both these factors created the possibility of a more in-depth look at the relationship between tsunami rehabilitation and the society that was at the receiving end of this aid.

This project, like many of the other studies carried out on tsunami rehabilitation was also interested in influencing policy debates. However, it adopted a novel strategy to achieve this objective. It sought to separate the research phase from that of generating ideas on policy implications. In other words, the researchers were given the space to carry out their tasks without being forced to come out with policy recommendations, at the end of their research. Most importantly, field studies were not policy driven but rather, each research study was built out of prior research knowledge and experience of the affected regions in which they were conducted, by researchers who had a long and committed scholarly and/or activist involvement in those regions. This strategy was adopted consciously because of the realisation that having policy compulsions right from the beginning might not only be a barrier to good research, but also would not produce good policy recommendations. This has often been the experience of research projects, unless they fall into the category of policy research. In the latter case, policy questions drive the research and policy recommendations are the primary objective of the research. My task in this project was to facilitate discussions about the intersections between research and policy and to tease out policy questions arising from the research, in conjunction with the researchers as well as policy makers in both Sri Lanka and Aceh.

The objective of this paper is to present the key policy questions that arise from the findings of the case studies of the project. As mentioned above, this is a comparative research project covering a number of locations in Sri Lanka and Aceh province in Indonesia. However, in raising issues for policy debates, the focus of this paper is

Sri Lanka. This does not mean the paper has ignored the findings of the Aceh component. The experiences in Aceh have been brought in for comparative purposes wherever relevant. Sometimes the Aceh findings are used to raise policy issues for Sri Lanka. It is hoped this strategy, while primarily focusing on Sri Lanka, will make some contribution to policy debates in Aceh as well.

The structure of the rest of the paper is as follows. The paper begins with a brief discussion on the links between research and policy. The main purpose is to clarify how this paper has approached this subject. The second section is an overall critique of the approach adopted by the bulk of tsunami rehabilitation projects in Sri Lanka. It covers a central policy instrument – Guiding Principles – which dominated the discussions within aid agencies. This is followed by the final section which focuses on some specific policy issues arising from research.

Links between Research and Policy-Making

Understanding the nature of the links between research and policy is an interesting area of work. There has been very little effort in Sri Lanka to explore how this process works. By and large the discussion centres on a blame game. When policy makers do not take notice of research findings, the usual reaction of the researchers is to blame the politicians or the bureaucrats for not having the capacity to understand the research findings, or for ignoring them because of other reasons like partisan political interests. The policy makers usually argue that some of the research is irrelevant and does not generate knowledge necessary for policymaking. In the case of politically contentious issues, researchers and policy makers often find themselves on opposing sides. Policy makers also find little time or inclination to digest research findings.

Those who have shown an interest in the issue conceptualise the link between research and policy as a linear process in which 'a set of research findings is shifted from the 'research sphere' over to the 'policy sphere', and then has some impact on policy-makers' decisions'.[1] This linear model is based on several assumptions: "First, the assumption that research influences policy in a one-way process (the linear model); second the assumption that there is a clear divide between researchers and policy makers (the two communities model); and third, the assumption that production of knowledge is confined to a set of specific findings (the positivist model)."[2]

The linear view of the links between research and policy has given rise to many activities. A dominant one is to focus on various means of communication. The question of links between research and policy has been reduced to one of communication and various strategies that are adopted to communicate the findings to the relevant policy makers.

Links between research and policy is much more chaotic and messy than implied by this model. The complexity of these links creates many entry points and strategies to influence policy. For example, research institutes influence policy through monitoring exercises, judicial activism, campaigning, lobbying, etc., in addition to more traditional publications and disseminations. Some of these strategies need not be directed only at policy makers. It can focus on the media with the objective of creating public opinion. Research itself can have number of objectives. Sometimes it can simply provide more in-depth information to policymakers. Or it can provide specific recommendations on the implementation process or make recommendations for reforms to the organisation implementing specific projects. Finally, research can question some of the fundamental concepts that underlie policies.

The exercise undertaken in this project belongs to the last category. Every project undertaken by aid agencies, or implementers

---

[1] Overseas Development Institute, Research and Policy in Development, "The RAPID Framework," accessed at http://www.odi.org.uk/Rapid/Tools/Toolkits/RAPID_Framework.html .

[2] *Ibid.*

funded by aid agencies, operates on the basis of a set of fundamental ideas. These ideas reflect a particular conceptualisation of the problem that they want to tackle, how these agencies understand the society that they are working in and formulation of a set of ideas as the solution. Since these fundamental ideas have been in operation for quite some time, they have become elements of a shared discourse within these agencies. These ideas find institutional expressions in the way these agencies are organised. Often they have become second nature to these organisations and therefore rarely questioned. The ICES research project was interested in questioning some of these fundamentals and unravelling this shared discourse. Therefore, the goal of this project's findings was not to provide specific guidance for implementing projects on the ground or making policy recommendations. Rather, it sought to raise questions about the basic assumptions that underlie the conceptualization of projects and policies.

Finally, it is important to note that the policy discussion in this paper is addressed to donors and other implementing agencies that depend on donor funding. Therefore, whenever the term 'policy makers' is used in the text, what is meant is policy makers within aid agencies or in implementing agencies that depend on foreign aid. This narrow focus allows for a more meaningful contribution within the limits of this project. In order to be successful, policy debates have to keep in mind the specificities of its audience. The manner in which policy debates are formulated and acted upon depends very much on the institutional framework within which these discussions take place. Each institutional framework sets limits within which policy questions are posed and acted upon. Often, it is important to understand the structure, logic, language, etc., of specific institutions in order to make useful policy recommendations. The methods used to convey messages will also depend on the specificity of organisations. For example, if the policy discussion is with the Sri Lankan state the type of research that needs to be done, the focus of research and strategies need to be adopted are quite different to when the conversation is with an aid agency. Taking these factors into account, this project decided to focus on aid agencies and implementing agencies closely associated with the former. Therefore this paper is addressed to them.

A Critique of the Overall Approach in Tsunami Rehabilitation

Responses to disasters by aid agencies are based on a particular understanding or an interpretation of the phenomenon.[3] The dominant approach is to treat disasters as isolated 'events' rather than a process characterised by the interrelationship between a natural phenomenon and society. When a disaster is treated as an event, the focus is on restoring what was destroyed (infrastructure, livelihoods, etc.) and doing it as soon as possible. What dominates is a discourse of emergency and restoration of the conditions that existed before the event. Of course, to do this not only are funds necessary, but they have to be spent as soon as possible.

In contrast to this there is a set of ideas on how to respond to disasters that has largely been initiated by those who came from the field of development. This approach focuses on the relationship between the natural phenomenon and society. It is much more interested in the links between conditions that existed in society prior to the natural event and disaster. It argues that the impact of the disaster is mediated through the structures of society that existed prior to disaster, and therefore there is a need to understand these conditions in planning a successful disaster management programme.

In this literature, a differentiation is made between 'hazards' and 'disasters'. The term hazard is used to identify the natural phenomenon. When hazards mediate through society we have disasters. Therefore, the term disaster is reserved for the analysis of the interaction between natural phenomena and society.

---

[3] See E. L. Quarantelli, ed., *What is a Disaster* (London/New York, Routledge, 1998) for a discussion on different theoretical approaches to disasters.

In responding to disasters, the focus of the latter approach is both emergency restoration and long-term mitigation. It does not ignore the emergency phase, but argues that even in the emergency phase the relationship between the natural phenomenon and society has to be taken into account. The fundamental objective is to improve the capacity of society to take care of disasters on a long term basis. This will involve many things other than construction.

A society-centred approach to disaster management will reveal that quite a few disasters in Sri Lanka are linked to land use and land ownership patterns. Most of the disasters in Sri Lanka are floods, droughts and associated phenomena like landslides. Many people who live and survive in locations such as low lying areas, non irrigated land and steep hilly areas suffer due to disasters. Many of these areas are unsuitable for human habitation. But some people are found in these locations due to the land ownership patterns of our society. For many poor people, these areas, unsuitable for living, are the only option they have. In fact, in urban areas one can see how market forces have literally pushed the poor people closer to water. The greater the demand for land by capital, the more likely it is that those who do not have capital will be pushed towards water. Hence, the land use and land ownership patterns have a bearing on how a natural disaster mediates through social structures. Land use and land ownership is only one, but a very key, aspect for understanding disasters in Sri Lanka. This analysis can be expanded bringing in many other dimensions of society that existed prior to the moment when the natural phenomenon struck the society.

This focus is essential for understanding the impact of the tsunami as well, although the scale of the phenomenon was such that it had an impact on a larger section of society. Quite a lot of people who lived near the coastline were affected by the tsunami. But the impact, as well as the capacity to recover, depended on the social positions of people. For example, in the case of the fishing community, it is the poorer sections of the community who live close to the vulnerable locations. They usually fish making use of smaller boats in shallow waters.

They have suffered significantly and they have difficulty in getting back on their feet. In some instances, the inflow of a large number of boats as tsunami assistance has made their conditions worse. There are now more fishing boats in shallow waters and lagoons competing for limited resources. Those who benefit from the surplus in fishing, such as fish *mudalalis,* or those who own multi-day boats that fish in the deep sea, might not even live right close to the sea. Even if they did, they would have had much more permanent and stronger houses that would have helped to minimise the effect of the tsunami.

If disasters are viewed in this manner, i.e., focusing on its linkages with societal conditions, disaster management gets closely linked with normal development issues. If we take its link to land use and land ownership, disaster management steps can range from resettlement, improvements and introducing preventive measures for those who are forced to live in difficult areas, various forms of support to improve the coping mechanisms of the people, etc. In addition, since anything to do with alienation of land and land settlement in Sri Lanka has a direct link with conflict issues, the disaster management strategies have to take into account the link between land and conflict as well.

The bulk of what went on in Sri Lanka in the name of tsunami reconstruction was dominated by the more traditional 'event focused' and 'emergency mode' responses. In addition this was backed by an unprecedented amount of funds, charity mentality and dominance of a large number of international agencies who would be in Sri Lanka only for a short period, and would not be there to face some of the problems that this approach might create. In other words, issues like long term partners, which are essential for sustainability of development projects, did not seem to bother those who came for tsunami rehabilitation. Many of the problems in tsunami rehabilitation arose due to the dominance of these fundamental ideas in rehabilitation.

Politics of Guiding Principles
Once the relief and rehabilitation got under way the government, donors and Non-Governmental Organisations (NGOs) agreed

o a set of Guiding Principles. These Guiding Principles reflected a framework or a set of normative principles which the implementers believed would lead to a more desirable outcome of the rehabilitation process. In key documents that reviewed the progress of rehabilitation here was always a section which assessed the ground situation in relaion to the Guiding Principles. The Guiding Principles included some of the following[4]:

1. Equity. The allocations of domestic and international donors should be guided by the identified needs and local priorities. There should be no discrimination on the basis of political, religious, ethnic or gender considerations. The recovery process should strengthen the peace process and build confidence. The reconstruction process should be sensitive to the impact on neighbouring, but unaffected communities.

2. Subsidiarity. The reconstruction activity should be designed and implemented at the lowest competent tier of the government to enable locally appropriate solutions, engagement of sub-national structures, capacity building and strengthening the different levels of governance and civil society.

3. Consultation. To secure the mid and long term needs of the victims, consultation, local decision making and full participation in reconstruction activities is essential. Interventions should respect local religion, culture, structures and customs.

---

[4] Information on these guidelines has been taken from the following three documents: Georg Frerks and Bert Klem, "Tsunami Response in Sri Lanka," Conflict Research Unit, Clingendael Institute/Disaster Studies, Wageningen University, March 2005. Joint report by the Government of Sri Lanka and Development Partners, "Post Tsunami Recovery and Reconstruction, Progress, Challenges, Way Forward" (December 2005); p. vii, Practical Action (ITDG), "Post-Tsunami Rehabilitation and Reconstruction: Compliance with Guiding Principles," Paper presented to the workshop 'Building back better-Are we on the right track?' held on 27th January 2006. There are slight variations in the versions given in each of the documents and thus my summary of some of the key ideas is an amalgamation.

4. Communication and transparency. There needs to be adequate communication and transparency in decision-making and implementation. This refers to policies, entitlements and procedures, as well as to resource use. All parties will adopt a policy of zero tolerance for corruption.

5. Reduce future vulnerabilities. Reconstruction should reduce future to natural hazards by adopting a multi-hazard risk approach.

6. Analysis of individual interventions. Interventions need to be assessed with their impact on prospects for peace and conflict, on gender, on the environment and on governance and human rights.

7. Debt relief. Revenues resulting from debt relief should demonstrably benefit the tsunami victims.

8. Co-ordination. Efforts need to be co-ordinated between all relevant stakeholders.

Many of the documents on the progress of implementation and the general thrust of the public debates suggest that there has been little compliance with these Guiding Principles. For example, a paper by an NGO presented to a workshop held in January 2006 to assess the progress of rehabilitation concluded that:

> Over the past year of implementation, there has been no strict monitoring of adherence to these principles. Therefore direct evidence of compliance, or its absence, is not available. However, there are reliable indications that concerted action has been wanting in this respect. The documents listed below contain information that clearly points towards poor compliance with Guiding Principles.[5]

---

[5] "Post-Tsunami Rehabilitation and Reconstruction: Compliance with Guiding Principles," *op. cit.*

The paper goes on to cite six key documents that have documented the progress of tsunami rehabilitation.

Much of the debate within the aid agencies and implementers
evolved around these guidelines in one way or another. It dominated
the debate and therefore set the terms of the discussion. Some NGOs,
depending on their interest, elaborated on some aspects of the Guiding
Principles. For example, Transparency International came out with an
elaboration on corruption.[6]

At first glance the Guiding Principles look very positive. They
include several ideas that can contribute to more desirable outcomes in
tsunami rehabilitation. However in order to understand what role these
Guiding Principles played in the politics of tsunami rehabilitation, let
us for a moment imagine what really has to happen if these principles
were taken seriously by those who implemented projects on tsunami
rehabilitation. If we take each element of the Guiding Principles one by
one, this would have meant efforts in the following broad directions or
at least some attempts in these directions:

- Equity – Interventions to undermine structures of power, institutions and practices that perpetuate discrimination

- Subsidiarity – Undermining the centralised state led by the
President and taking action to strengthen sub-district units politically.

- Consultation – Setting up organisations of the victims themselves and involving them in planning.

- Communications and Transparency – Ensuring that victims
or their organisations have the information and can act upon it.

---

[5] See, Transparency International Sri Lanka, *Preventing Corruption in Post-Tsunami
Relief & Reconstruction Operations: Lessons and Implications for Sri Lanka*, A briefing
note (Colombo: April 2005).

- Reduce future vulnerability – Move way from a focus of restoring to what existed before the tsunami, to a long-term strategy

- Analysis of the impact on conflict, gender and environment – Once again analysis and interventions on structures of power, institutions and practices that sustain the conflict, patriarchy and destroy the environment.

- Debt relief – Interventions at the level of the Finance Ministry

- Co-ordination – Sometimes this amounts to only sharing information. But co-ordination in terms of working on the basis of common plans and objectives demands that each agency ignores their own mandate and interest and joins together with others.

If the Guiding Principles are viewed in this manner, it is clear what is implied in them is a difficult political project. This political project has to tackle the centralised state, structures of power in society organise the victims, ensure that they get information that they can act upon, have a long-term view of disaster management and reform how aid agencies work. In short, it is difficult to expect any of the objective set out in the Guiding Principles to be achieved without a serious political commitment to transforming existing structures of power. What happened was far, far away from this political project.

Then the question arises as to what does this type of exercise of establishing principles actually mean within the politics of the aid industry. The origin of these ideas can be traced back to two fields within the aid industry – development and governance. The Guiding Principles are a collection of ideas generated from these two fields that contribute towards the legitimisation of the activities of aid dependent agencies. They help these agencies to be politically correct.

However, the very process of translating these potentially radical ideas into principles, depoliticises them. What happens is that a set of ideas which are deeply political, and whose fulfilment demands a significant political intervention, get transferred into a set of principles

hich everybody round the table can agree on. In other words, the po-
:ical content of what they mean has been ignored, and the politics of
)cial transformation embedded in these ideas are converted into some-
1ing else. Once these ideas are depoliticised through such exercises, in-
ead of political engagement what we have are reports and discussions
)out compliance. This is followed by constant repetition of these slo-
ans without the political content. In the end it gives the impression
1at these agencies are interested in fundamental reforms without ac-
1ally having the political commitment to carry them out.

The pernicious effect of this type of exercise is much wider
/hen these discourses begin to dominate the public debate. This is
/hat happened during tsunami rehabilitation. Institutionalisation of
hese discourses, supported through funding, crowded out other types
)f critical thinking. Challenging them is a major task of critical re-
earch.

Research Findings Relevant for Policy Discussions

The rest of the paper focuses on some of the key findings of re-
;earch that need to be considered by the implementers. In discussing
hese findings Guiding Principles, which was the principle manage-
nent tool that tried to set a normative framework, will be referred to
wherever relevant.

Avalanche of Goodwill – How to Manage it?

The tsunami created a global level response from a large number of sources. Probably what we saw was a new phenomenon of charity in a globalised world. A variety of transnational structures supported
by a globalised media helped to bring about this globalised response.

On one hand, this is a reflection of a sense of goodwill. To this
extent it had a positive side. But on the other side, it also brought in a
plethora of agencies and individuals with very different worldviews and
experiences into the recipient society. Many came without much of an
understanding of the ground situation. Very soon, they began to face
numerous difficulties and the blame was on the recipient society. Once

they faced challenges due to the nature of the society that they we
working in the instinct was to blame the victims who were ungratef
and had not created the proper conditions so that the funds that we
so generously given could be utilised.

This contrasts sharply with the attitude of developmer
projects. In development projects, there is a recognition that you a
working in societies very different to those from where the funds orig
nate. This recognition creates the need to understand the recipient sc
ciety in all its complications. These complications are seen a
challenges, and overcoming these difficulties is an essential part of de
velopment. Therefore, the attitude of a good development practionee
is very different from those engaged in emergency humanitarian assis
tance.

There is no doubt that the massive influx of international orga
nisations and individuals created a complex set of problems. Some o
these issues are about the perceptions that this type of an influx create
in the host society. Others are related to more practical issues of effec
tiveness of interventions, accountability if these interventions ge
wrong.

The magic word that all donors dealing with this massive influ
of aid resorted to was 'co-ordination'. This was introduced as part o
the Guiding Principles for implementation. But this is to reduce a se
rious political issue about interests of the agencies, how they are struc
tured, to the language of management.[7] Many of the agencies whc
undertook tsunami reconstruction have their mandates and hierarchica
structures that link them to their headquarters, usually located in west
ern capitals. Much of their planning goes on within these organisation
al structures. To shift these mandates in such a way so that they can
have a common programme with a host of other agencies who have
their own mandates and structures is almost impossible, unless deci-

---

[7] For a very good critique of the notion of coordination in post-tsunami rehabilitation
in Sri Lanka, see Jock Stirrat, "Competitive Humanitarianism: Relief and the Tsunami in Sri Lanka," *Anthropology Today* 22, no. 5 (2006): 11-16.

ons to do so are taken at the highest policy making level. Therefore
ie kind of co-ordination that Sri Lanka needed, which amounted to a
tuation where all actors worked within a framework that would have
enefited Sri Lanka, was almost impossible to achieve.

The introduction of a notion of 'co-ordination' precluded any
ype of discussion about the internal workings of these organisations.
asically, it is assumed that there is no problem with them and they
eed not change. What needs to happen is to 'co-ordinate' with other
gencies, whose mandates also remain intact.

The negative impact of massive external interventions has been
n issue that has been raised in many other situations - e.g. Rwanda,
fghanistan, Bosnia-Herzegovina - where there were significant inter-
entions by external actors. But very little has happened to tackle the
iegative impact caused by the very presence of such a large number of
gencies. Tsunami rehabilitation in Sri Lanka was one more case in this
tring of policy failures.

How to Deal with the State?

Findings in both locations – Aceh and Sri Lanka – where research was carried out, shows the importance of understanding the nature of the state and its behaviour in planning interventions.
Sometimes it is necessary to keep mentioning this issue because in an
environment of liberal orthodoxy the state is always seen only as a
problem. There seems to be little effort to understand the nature of the
state in societies where donor-supported interventions are taking place.
This question becomes even more difficult when both the character
and the behaviour of states varies greatly in different countries. In
short, states of the global South cannot be understood with a broad
brush. In the history of these countries there are instances where disasters triggered fundamental structural changes. This possibility becomes
even more viable in countries where there have been ongoing conflicts.
The behaviour of the state is determined by many of these considerations.

The initial response of the Sri Lankan state towards tsunami rehabilitation was guided by the current orthodoxy of the state sector playing a minimal role of setting the overall framework, while handing over the implementation to private actors. In other words, the Sri Lankan government pretty much privatised tsunami rehabilitation. The government set up what was called the Task Force for Rebuilding the Nation (TAFREN) which answered directly to the President. In keeping with the prevailing ideology, the members of the board of TAFREN came from the private sector. The private sector was also brought in to set up the implementation mechanisms of TAFREN. The highly centralised state of Sri Lanka, where the President enjoys so much power, cleared the way for these decisions.

In this privatisation process of tsunami rehabilitation International Non Governmental Organisations (INGOs), along with their local partners, became principle actors. In most instances state agencies only played a peripheral role by providing information. When the state came back into the picture the language of disaster management dominated,[8] although many of the activities that constituted rehabilitation were normally carried out by line agencies or provincial councils.

The fact that Sri Lanka depended on external funding played a significant role for this particular behaviour of the state. The massive flow of funding had to be co-ordinated by centralised institutions. Its usage had to be organised according to the orthodoxy that prevailed. Implementation had to be privatised. Finally, since the funds were meant to be for recovery from a disaster, state institutions dealing with development was largely kept out. Thus, foreign funding was a dominant factor that determined the behaviour of the state.[9]

Subsequent developments showed the initial model, which lasted throughout 2005, the first year of tsunami rehabilitation, was

---

[8] For an extended discussion on disaster management and disaster risk management, see Vivian Choi, in this issue.

[9] A similar experience was had when the Sri Lankan state implemented the Accelerated Mahaweli Development Programme.

y much an idea of President Kumaranatunge. As soon as the Presi-
it changed after the December 2005 elections, the attitude towards a
s interventionist role of the state changed. With the conversion of
FREN into the Reconstruction and Development Agency (RADA),
der President Rajapaksa, the state became more interventionist.
wever, the interventionist role was most pronounced at the central
el. The President wanted a greater degree of control at the central
el, while what happened in the periphery remained almost the same.

The behaviour of the Sri Lankan state had an impact on how
affected population perceived the responsiveness of the state. This is
rne out by some of the surveys carried out among the tsunami survi-
:s with regard to the responsiveness of different actors:

> A significant difference between India and Sri Lanka in the
> tsunami relief effort was the role of government. In India,
> where the government played a critical role in coordinating
> the rescue and relief efforts, the affected families reported satis-
> faction with the visible and tireless district level administration
> who provided and co-ordinated relief. In fact the government
> was ranked as the number one provider by the affected people
> on all dimensions of relief services.[10]

The marked absence of state responsibility in the rehabilitation
cess in the case of Sri Lanka can create difficulties in the future. For
mple, many private actors have been responsible for building hous-
complexes. But who will be responsible for the maintenance of
nmon amenities in these complexes in future? There is a long histo-
of INGOs implementing various projects and vanishing from the
ne, leaving the host country to face problems in the future. Usually
velopment projects tackle this issue right from the start by undertak-
; these projects in partnership with local actors, usually the state.
ice the project ends the partners are responsible for follow up work.

---

ritz Institute, *Lessons from tsunami: Top Line Findings* (Fritz Institute, 2005).

Generally those engaged in emergency responses do not tackle these
sues. The emergency nature of the response helps external agencies
ignore this type of problem. What really happened in the case of t
tsunami was those who came for emergency assistance ended up ir
plementing development projects. To complicate matters, the sta
played a minimal role in implementation. It is quite possible that tł
model will result in many problems in the future without anybody r
sponsible to take care of them.

In the case of Aceh, the relationship with the Indonesian sta
was mediated by the conflict that prevailed. Before the tsunami, Acc
was a highly militarised province. The Indonesian state viewed this ar
primarily through the lens of tackling an insurgency through milita
means and maintaining the coherence of the Indonesian state. The tsι
nami and the peace process that followed forced the Indonesian state t
rethink this behaviour. The Indonesian state responded positively t
the demand for autonomy, but at the same time began to take measure
that would maintain the cohesiveness of the Indonesian state.

The important policy issue that needs to be noted here is that
state's behaviour is extremely complex and often determined by inter
nal political compulsions. Normally, agencies deal with this all impor
tant entity through managerial logic. Terms like capacity building
improving accountability, etc., are utilised for this purpose. Concept
in the Guiding Principles such as subsidiarity, consultation and trans
parency are no substitute for a serious analysis of the state and identify
ing what can and cannot be done with it. The real problem then is tha
there has been little effort made to look at the complexity surrounding
the state. A much more nuanced understanding of the state can provide
a better guidance to policy.

What is the Unit of Intervention or Planning?

At the initial stages of humanitarian intervention, the focus was
on the people who suffered from the event. This ranged from individuals to entire villages. During the rehabilitation phase, especially when
projects went into areas like construction of houses, the larger social

ınit became much more relevant. The interesting question is how implementers identified this social unit in their planning process.

Research in Aceh clearly shows the importance of this issue. Saiful Mahdi's study, for example, points out the importance of *gampong*,
or the smallest unit of communities, both for facing difficulties that the
tsunami brought about and for rebuilding.[11] According to these findings, the social capital associated with this unit has been critical for the
resilience of those who suffered from the tsunami. The paper contrasts
this unit with what is called *desa* or *keleuran* , which was brought about
as a result of legislation originating from Jakarta. This unit does not
have similar resonances in Aceh. Therefore *gampong* is the more appropriate unit for planning interventions.

The Sri Lankan equivalent of *desa* is the *grama niladhari* (GN)
division, the lowest administrative unit. The official heading this unit is
the *grama niladhari*. The term *niladhari* or 'official' denotes that he is
part of the public administration. This unit was established in the mid
fifties and by now has become an entrenched part of the public administrative system.[12] However, within a GN division or sometimes cutting
across GN divisions, are other social units, popularly called villages.

During tsunami rehabilitation of Sri Lanka the *grama niladhari*
figured prominently as a source of information. He or she also entered
into the picture during the selection of beneficiaries. But the GN division did not figure that much as a unit of planning. In fact, in the case
of Sri Lanka there seems to have been very little discussion about the
unit of planning. On the one hand, where the pure humanitarian logic
dominated, being affected by the tsunami was enough to make that individual or household, the unit of planning. On the other hand, many
thought using the word 'community' solved the problem. However,

---

[11] Saiful Mahdi, in this issue.
[12] For a further discussion on this administrative unit, see Pradeep Jeganathan, in this
issue.

very rarely was this term 'community' given any sociological meaning or questions posed about its relevance for planning.[13]

What is interesting is that the Guiding Principles did not provide any opening to tackle these types of issues. However, these types of issues have been debated in Sri Lanka in the context of development projects. For example, during the resettlement of people in dry zone settlements like Gal Oya and Mahaweli, social units within which people lived before resettlement became an important issue for planning resettlement. There were even discussions about the importance of caste in Sri Lankan villages and the need to take this into account in future settlements. The important issue here is not whether we agree with these conclusions or not, but the fact that development interventions provided spaces for these types of debates which could go into detail about the nature of the society that one was dealing with. Unfortunately, the importation of concepts that donors repeat all over the world, as reflected in the Guiding Principles, not only precluded such debates but also prevented learning from past experiences of development planners in Sri Lanka.

How to Deal with Land Tenure Issues?

Tsunami rehabilitation soon became a large-scale housing project. Since houses have to build on land, projects had to deal with land tenure issues. This is not unusual for development projects. Development projects dealing with housing often have to tackle land tenure issues. In countries like Sri Lanka and Indonesia, the land tenure picture is not that simple. There can be several forms of it and it can change from area to area. In addition, one cannot determine the land tenure picture only by looking at legal documents. Field work is necessary. For example, the term 'hidden tenancies' is a common term in agrarian studies. This denotes a situation where the actually existing te-

---

[13] For a critique of the notion of 'community' constructed by aid agencies, see Pradeep Jeganathan, in this issue.

nancy, that can be discerned by studying cultivation practices, is quite different from what appears on official documents.

Therefore it is no surprise that the issue of identifying land became a gigantic and complex question for those involved in housing projects. In Sri Lanka, it got even more complicated by the government's decision to implement a uniform buffer zone throughout the island.[14] There is also evidence that the interest of capital, especially tourism, played a role in this decision.

This experience shows that land tenure issues are a central underlying factor of housing projects. Questions of land tenure are always accompanied by various types of struggles and conflicts. For development projects that focus on housing, this is nothing new. Understanding land tenure patterns in the locality where projects are implemented, taking steps to sort out land tenure issues and leaving enough time for all this is common in development projects. For humanitarian organisations whose aim is restoring what existed before as soon as possible, these issues become a problem. They do not have time to look at the complexity of society. Societies are seen only through the event which brought them into the picture. Hence, land tenure becomes a problem that somebody else has to resolve. This approach, which treats disasters as an event isolated from social structures, is not only incapable of tackling these issues, but can leave behind many problems.

Land was one issue where tsunami rehabilitation could have been used to bring about structural reform. This was regarding people who either rented houses or illegally squatted on land. Due to land hunger in the densely populated coastal areas, there are settlements where people have occupied land illegally. These are mostly poor people who do not have capital to own land in these locations. It was possible to make use of rehabilitation projects to provide some respite to such communities.

---

[14] For a further discussion, albeit from different angles, of some of the consequences resulting from the institution of this buffer zone, see papers by Malathi de Alwis and Jennifer Hyndman, in this issue.

However, what dominated in tsunami rehabilitation was the notion of non-discrimination as stated in the Guiding Principles. This means treating those who were affected equally. In other words, there was no space for positive discrimination on behalf of the poor. True, the problem of those who lived in rented houses came up. But it was resolved through providing some monetary relief, rather than finding a solution of a structural nature. There is very little specific knowledge about how the projects dealt with those who occupied land illegally.

The experience with the land issue, which was intimately linked to housing, raises a whole host of policy issues. Some of them have been dealt in development projects that have focused on housing. Hence there is a certain amount of knowledge out there that can be utilised. However, in order to benefit from this knowledge, rehabilitation programmes have to get away from the 'emergency mode' and preoccupation with restoring what existed before as soon as possible. The usual counter argument in this regard is that these agencies are dealing with emergencies and cannot operate like development projects. But if we think for a moment and realise that housing projects are still going on, i.e. close to four years after the tsunami, what we are dealing with are really development projects and not emergency rehabilitation. Therefore lessons that have been learnt from development projects are valid.

How do you Generate an Adequate Knowledge Base of the Society in which you are Intervening?

The issues raised above such as how do we understand the nature of the state within which projects operate, what is the relevant social unit for intervention and complexities of land tenure, raises the general question of how to generate an adequate knowledge base for humanitarian interventions. Some of this knowledge which gives basic data about the extent of damage and number of victims has to be generated anew. It also has to be done as quickly as possible. This part of knowledge generation seems to have been carried out quite satisfactorily, both in Sri Lanka and Aceh.

Problems arise when we go into the rehabilitation phase. It is at this point that the type of issues that I have raised above come into play. It demands more in-depth understanding of society. At this stage here is no escape from the fact that projects have to make use of the existing knowledge base of societies and build on that wherever necessary. This has not been happening. Instead there is a dubious process of knowledge generation that can be questioned at both theoretical and methodological levels.[15] This has also been one of the major complaints of the research community.

It is from a more in-depth understanding of a society that principles that can guide rehabilitation programmes can be drawn. In contrast to repeating slogans that donor agencies market around the world, what is needed is an understanding of the specificities of the society where agencies are working. Humanitarian agencies and donor groups must get away from the temptation to use huge generalisations to understand the Global South. Concepts such as 'Third World' or 'Failed States' simply ignore complexities within these societies and are intellectually sloppy.

This brings me to the question about who should be responsible for generating such knowledge. At the very outset, let me state that this is not a question about those who are living in the Global South versus those who are outside. It is also not a question of white *vs* other colours. It is more a question of who is qualified and who has the experience.

In my own career of working as a consultant for numerous donor funded projects, for more than fifteen years, I have had very different experiences of the quality of external consultants that I have worked with. When I began my career, mostly in development

---

[15] For a discussion of some of these theoretical and methodological problems with reference to knowledge-generating mechanisms such as 'participatory rural appraisal', see Malathi de Alwis, in this issue. For a discussion regarding projects focusing on gender issues and INGO responses to Islam, in Aceh, see Jacqueline Siapno, in this issue.

projects, most of my colleagues were either people who had either done their post-graduate work or some other form of studies, in Sri Lanka. Some others had a development studies background or were grounded in area studies. Many of them were aware of scholarly and political debates in Sri Lanka and were familiar with the academic literature on Sri Lanka. Many of them had also carried out field work in local languages. Their understanding of social science methodologies was also sound.

There is a striking difference between these early experiences of mine and more recent ones, especially after 'conflict transformation' became a buzz word among the donors. For example, it is not unusual to find consultants talking and writing about conflict without using the word politics. If in the early period there was space for taking into account histories, social structures and politics of the specific societies, in a situation like a civil war where this understanding is essential, there are many consultants who not only do not have this background, but also lack training even to pick it up. Let me make it clear here, I am not making a sweeping generalisation, but a specific comment about individual experts. This means I have also had the privilege of working with individuals who know the politics of this country or have the necessary training to pick it up quickly. The major point I want to make is that this experience is much rarer now than when I worked in development projects.

This personal digression was necessary to re-emphasise the point that was made earlier – how global blueprints are substituting for context-based knowledge. Many of the consultants who go round the world work with these global blueprints. Many projects also work within these blueprints. In such contexts, there is no effort made to generate context-specific knowledge or to make use of those who have an extensive expertise in the peoples and regions where these projects are being implemented. The result is bad projects as well as bad policy.

How to Tackle Established Structures of Power?

Tsunami rehabilitation has raised many issues about its implementation process. Many of these complaints are about equity issues.
In short, there is the usual complaint about the benefits being captured
by influential sections of society. The other side of the coin is some of
the tsunami affected have been left out. Some of these complaints are
merely based on perceptions. But others have validity.

These debates raise fundamental issues about the interactions
between the flow of benefits and power structures in society. For those
who would argue that disasters should be seen as an interaction between an event and society, the importance of this issue is self evident.
Rather than engaging in endless debates or doing surveys to find
whether these complaints are true or not, what we need to ask is
whether project interventions have adequate tools to subvert power
structures so that those who should benefit from the interventions actually get their due.

In trying to deal with these issues, two concepts dominate the
current discussion – 'community participation' and the establishment
of procedures for accountability. Both these ideas were reflected in the
Guiding Principles. The first idea comes from development practice
and the second from the discourse of good governance. However both
might be ploys to ignore the central issue that needs to be looked at –
the question of power structures. The notion of 'community' has been
used for a long time to ignore the presence of caste, class, gender domination, ethnic discrimination, etc. Pradeep Jeganathan, in this issue,
has critically examined the notion of 'community' constructed by aid
agencies. Similarly, the notion of 'good governance,' addressed by Malathi de Alwis, in this issue, is an attempt to develop a rule-based liberal
democratic state. Once again, the issue of power struggles within the
state are ignored. Many of the specific ideas that dominate project implementation can be traced to these concepts.

An interesting aspect demonstrated by studies undertaken in
this project is the element of self organisation among the tsunami-

affected when they faced injustice.[16] This has taken various forms an occurred outside the development project structures. They made use c existing subaltern spaces in society for their protests and linked up wit whoever was useful for their purposes; it could be political parties, rel gious institutions, media, etc. This is not participation of the commu nity as dictated by projects, but instances of self organisation. They ar authentic voices arising from perceived or real grievances.

Participatory discourses of donor supported projects not onl stay far away from these expressions of the people, but try to substitut an alternative which they can control. Therefore, it is never able to sa tisfy the needs of the community. There have been many writings anc many criticisms of participation in academic literature.[17] But the ver fact these ideas were repeated once again in the context of tsunami re habilitation demonstrates how little aid policies have moved over th years.

Participation as practiced by donor supported projects hardly leads to challenging the structures of power that underlie inequity in the flow of foreign aid. Once again it is a discourse that undermines the political action which is needed to achieve objectives of equity. It is time that policy debates subject the discourse of participation to a radi-cal critique.

The Debate on Tsunami
Internally Displaced Persons (IDPs) and Conflict IDPs

In both Aceh and Sri Lanka, there was a discussion, among do-nors and project implementing organizations, on whether or not to in-clude people affected by conflicts, in the tsunami assistance programmes. In both locations, pressure and in the case of Aceh, agita-tion, was necessary to make the voices of conflict IDPs heard.[18] Fortu-nately, this issue was resolved to a certain extent. However, the very fact

---

[16] See Malathi de Alwis and Eva-Lotta Hedman, in this issue.

[17] For a discussion of this literature and various debates, see Malathi de Alwis, in this issue.

[18] See Eva-Lotta Hedman, in this issue.

his type of lobbying and pressure was necessary to recognise the simple act that ignoring people affected by conflicts would lead to more prob-ems says a lot about the institutional logic of humanitarian agencies.

The Guiding Principles mention that the interventions need to be assessed for their impact on peace and conflict, gender, the envi-onment, governance and human rights. Hence, conflict became one more thing in the laundry list that is taken round the world by donors. This is a formulation that never leads to any serious analysis of links between rehabilitation and the politics of peace.

A closer look at the behaviour of aid agencies demonstrates that what determines what they do is their institutional logic. Although here is an elaborate mechanism to legitimise the activities of aid agen-cies on the basis of benefits to the recipient society, impact of projects, and various conceptual frameworks, what actually determines beha-viour is the institutional logic of the agencies.[19] For example, if an agency identifies itself as a humanitarian organisation, receives funding to take care of those affected by a particular event, and has an internal logic to operate on this basis, this is what will prevail irrespective of how this approach will impact on the recipient society.

Many agencies will argue that this is a result of the specialisa-tion of various agencies and this is necessary for them to be effective. But the problem arises when there is a contradiction between this spe-cialisation of the organisation and resulting institutional logic and the demands of societies in which these agencies begin to operate. Al-though there have been discussions about adaptable organisational structures, and even building organisations based on chaos theory which is more suitable for dealing with complexities of societies, very little progress has been made on this front. In the complex world in which aid agencies work, such institutional adaptations are essential.

---

[19] For a very useful contextualization and historicization of the institutional logic of Canadian International Development Agency (CIDA), see Jennifer Hyndman, in this issue.

Sunil Bastian is a political economist and development consu‑
tant. He has published widely on issues of devolution, developmer
and aid. His most recent publication is *Politics of Foreign Aid in S.
Lanka, Promoting Markets and Supporting Peace (2007).*